HALF-TRACK

A HISTORY OF AMERICAN SEMI-TRACKED VEHICLES

by
R.P. Hunnicutt

Line Drawings
by
Michael Duplessis

Color Drawing
by
Uwe Feist

FOREWORD
by
Major General Oscar C. Decker, U.S. Army (Retired)

E P B M
ECHO POINT BOOKS & MEDIA, LLC

Published by Echo Point Books & Media
Brattleboro, Vermont
www.EchoPointBooks.com

Copyright © 2001, 2015 R. P. Hunnicutt
ISBN: 978-1-62654-860-2

Cover image by Uwe Feist

Cover design by Adrienne Núñez,
Echo Point Books & Media

Editorial and proofreading assistance by Christine Schultz,
Echo Point Books & Media

Printed and bound in the United States of America

CONTENTS

Foreword... 5
Introduction.. 6

PART I HALF-TRACK DEVELOPMENT PRIOR TO WORLD WAR II............. 7

Early Development... 9
Half-Track Cars... 12
Half-Track Trucks.. 15

PART II HALF-TRACKS FOR WORLD WAR II................................. 23

Armored Half-Tracks.. 25
Half-Track Cars and Personnel Carriers.. 29
Mortar Carriers.. 80
Tank Destroyers... 97
Self-Propelled Artillery... 112
Antiaircraft Vehicles... 122
Specialized Half-Track Applications.. 158
Three-Quarter Track Vehicles... 161
Half-Track Amphibians.. 172
Lightweight Half-Tracks.. 174

PART III THE HALF-TRACK GOES TO WAR.............................. 175

Half-Track Operations During World War II................................... 177
Korea... 195
Half-Tracks in Foreign Service... 199

PART IV REFERENCE DATA... 201

Color Section... 203
Half-Track Acceptances.. 208
Vehicle Data Sheets... 209
Weapon Data Sheets... 227
References and Selected Bibliography.. 238
Index... 239

ACKNOWLEDGEMENTS

First, I would like to express my thanks to Major General Oscar C. Decker who kindly agreed to write the Foreword for this book.

Also, I am particularly indebted to David R. Haugh who allowed me to use much of the research material for his excellent book "U.S. Half-Tracks, Their Design and Development".

Fred Pernell, formerly with the National Archives in Washington D. C. was a great help in finding my way through the mass of material available in the Archives. In fact, he tracked down many of the key documents himself.

At the Patton Museum, John Purdy, Charles Lemons, and Candace Fuller located many valuable items, much of which came from the collection of my old friend the late Colonel Robert J. Icks. His collection is now in the Museum Library. Also at Fort Knox, Bill Hansen, Chief Librarian at the Armor School, was most helpful in finding material on armor operations in the Philippines during the early days of World War II.

Jon Clemens at Armor Magazine found numerous half-track photographs which are included in the book.

The United States Marine Corps Museum at Quantico, Virginia was the source of many photographs showing Marine Corps half-tracks in operation. Ordnance Specialist Dieter Stenger was particularly helpful.

The Ordnance Museum and Library at Aberdeen Proving Ground provided valuable information from their collection. I am particularly grateful to the Director, Dr. William F. Atwater, and to Alan Killinger for their help. Dr. Peter S. Kindsvatter also located some very useful information.

Most of the material on the International Harvester half-tracks would not have been available without the help of Greg Lennes and Julia Brunni of Navistar International Transportation Corporation. A photograph of the M2 half-track car also was received from Mr. Phil Rombo of Volvo Truck of North America.

Michael Duplessis prepared the five view drawings and did much of the research on the details of the various vehicles. As usual, the color drawing of the 75mm gun motor carriage M3 was the work of Uwe Feist.

Many of the photographs would not have been available without the help of Steve Zaloga. His numerous works on half-track vehicles were a valuable source of information on many of the field modifications made by the troops.

Fred Crismon was the source for both data and photographs of some of the more obscure vehicles. In this regard, his collection is unmatched.

Mike Green also located numerous photographs for the book.

A number of problems were solved during the research program by examining the half-tracks in the collection of my friend Jacques Littlefield. Roy Hamilton was particularly helpful in sorting out the details of the half-track suspension development.

Special thanks go to Randy Denna and the crew at Ritz Camera for helping solve some of the problems with the photographs.

FOREWORD
by
Major General Oscar C. Decker, U.S. Army (Retired)

I know of no one more dedicated or more qualified to tell the story of U. S. armored vehicles than Dick Hunnicutt. As we have come to expect from his outstanding books covering the history of U. S. armored vehicles, he has done it again. The detail in this story of the half-track carefully describes the process of taking the results of slow early development (pre-1040s) and then hurrying one of the first armored "horses" through development, production and many model changes. It demonstrates again the painstaking research for which Dick is famous.

As I look back to the days of the venerable half-track and my introduction to it as an enlisted soldier in Armor during World War II, it brings mixed memories. In normal terrain, it provided excellent mobility, but in mud the wheels tended to dig themselves in so that the tracks literally seemed to drive the wheels further into the ground. Mounting the .50 caliber machine gun made it a formidable vehicle against ground troops and thin skinned vehicles. I recall one particular day outside a small town north of Munich, Germany when I saw it come into play rapidly and effectively. Our tank battalion was temporarily halted and a number of us were out on the ground. Suddenly, a German soldier, apparently wanting to be a hero, opened fire from a nearby building. Before any tank could react, the Sergeant on the maintenance half-track swung his .50 caliber machine gun around and cut out the window frame and the resistance. After I was commissioned in 1951, my memories of the half-track are on border duty with the 82nd Reconnaissance Battalion where we used it as a communications relay station. The engine could run 24 hours a day with minimal maintenance and excellent reliability, a great comfort to us.

Having had several assignments at the Army Tank Automotive Command, it occurs to me that the apparent rapid development and acquisition process that was demonstrated in the early 1940s history of the half-track might be instructive to us today as we continually try to improve and shorten the acquisition cycle. Ideas from the armor soldier seem to have been quickly studied and incorporated. As the history unfolds, in some cases they were put into place by our ever ingenious maintenance soldiers even before they were approved. Components were directed to multiple manufacturers while today we seem to be reluctant to give that type of specific direction. In another interesting approach to development, Colonel Hinds is described as acting on an idea from two of his soldiers for replacing the fixed idler with a coil spring. He paid for the parts to test it and took it to Major General Harmon who approved and had it installed on the half-tracks slated for North Africa.

The book skillfully takes us from the meager development effort in the pre-1940s, which was primarily modifying wheeled vehicles with a partial track, to the formal armored half-tracks which most World War II and early postwar veterans know so well. It traces the lineage as personnel carriers, reconnaissance vehicles, artillery prime movers, self-propelled weapon carriers including mortars, tank destroying guns and antiaircraft weapons, command and engineer vehicles, ambulances, and communication vehicles. As described, the half-track perhaps should also be credited with being one of the first "common chassis" vehicles which we like to talk about today.

This is another important book to add to the libraries of current and future generations to show U. S. armored vehicle development.

INTRODUCTION

The family of half-track vehicles that served the United States and its allies during World War II was descended from the early Citroen-Kegresse half-track purchased from France in 1931. Although the semi-tracked configuration had been applied to numerous earlier vehicles, most of them were slow and not capable of rapid movement on roads. Tests of the Citroen-Kegresse P17 resulted in the initiation of a development program for a series of half-track cars and trucks prior to World War II. A few of these were produced in small numbers and went into service with the troops as personnel carriers or prime movers for light field artillery.

In 1938, an M2A1 scout car was modified by replacing the rear wheels with a tracked suspension from one of the half-track trucks. The converted vehicle was designated as the half-track personnel carrier T7. Although it was underpowered, the test program of the T7 provided valuable information for further development. It also was the first armored half-track in the U. S. Army. Based upon the experience with the T7, a new vehicle was designed during 1939 and designated as the half-track scout car T14. After modification, this vehicle was standardized as the half-track car M2 and it provided the basis for all of the half-tracks produced during World War II. In addition to the half-track cars, they included personnel carriers, mortar carriers, tank destroyers, self-propelled artillery, and other specialized vehicles based upon the standard half-track chassis. Many of these were expedient solutions until a properly designed vehicle was available for a particular application. In the early days of the war, the half-track often was the only vehicle in production that could be utilized. Some of the expedient vehicles were replaced by full track vehicles as soon as they became available. Others were so successful that they remained in production.

The half-track in its various modifications served from the first days to the end of World War II. Its early production made it available for the defense of the Philippines after the attack on Pearl Harbor. It continued to serve in all theaters of operation during World War II until the Japanese surrender in 1945 and a few versions were not declared obsolete until long after the end of the Korean War.

Although this volume is primarily a development history, a few photographs are included showing its service both in training and in combat. However it is not an operational history. This part is intended to illustrate the effectiveness of the vehicles and some of the modifications made by the troops in the field. Combat experience resulted in some modifications being standardized and applied to a number of vehicles on a production basis. The M16A1 and M16A2 multiple gun motor carriages are examples of such improved vehicles resulting from battle experience during the Korean War. In Japan, the M15A1 combination gun motor carriage was rearmed with a single 40mm gun because of the shortage of 37mm ammunition during the early fighting in Korea. The modified vehicle was then designated as the 40mm gun motor carriage M34 and classified as Limited Standard.

After Korea, the half-track rapidly disappeared from the U. S. Army, but it remained on active service in numerous foreign armies. Israel in particular, utilized large numbers of half-tracks. Many of these were modified, rearmed with new weapons and powered by new engines to improve their performance. Even today, these World War II half-tracks in modified form are still employed in some foreign armies more than half a century after their introduction.

The specifications for many of the half-tracks and some of their weapons are outlined in Part IV. Some production data also are included in this reference section.

As usual, many questions are raised during the research on a project such as this and many of them are never satisfactorily answered. Frequently, some of the answers appear after the book is published. The author would be grateful for any additional information that would help complete the story.

R .P. Hunnicutt
Granite Bay, California
October 2000

PART I

HALF-TRACK DEVELOPMENT
PRIOR TO WORLD WAR II

Above are two views of the 3 ton Garford-Holt half-track designed for the Quartermaster Corps. It was photographed during June 1918. The 3 ton Packard truck fitted with the Holt track rear suspension appears below at the right.

EARLY DEVELOPMENT

In the United States, most of the early commercial tractors utilized the semi-tracked configuration. This design carried the main weight of the vehicle on a tracked suspension to lower the ground pressure and used separate wheels for steering. In many cases, track braking also was used for greater maneuverability. These vehicles used in construction and agricultural applications were heavy and slow moving. Eventually, the steering wheels were discarded and track steering alone was used to guide the vehicle. Even the early Mark I British tanks were provided with two rear wheels for steering purposes. These wheels were discarded early in the program in favor of track steering alone.

In addition to the use of heavy artillery tractors during World War I, the semi-tracked configuration was considered to be a possible solution to the problem of improving the cross-country mobility of trucks and other light vehicles. For evaluation purposes, small Holt tractor suspensions were installed to replace the rear driving wheels on some trucks. One of these was the conversion of the 3 ton Garford truck tested during 1918. The front wheels were not powered on this vehicle. Although it was slow, with a maximum speed of

only about seven miles per hour, this installation reflected the half-track configuration that was to be widely developed in the future.

During this same period, the Holt Tractor Company modified a 3 ton Packard truck by replacing the rear wheels with the same tracked suspension used on the Garford. Unfortunately, it too was extremely slow.

Although it was not completed until 1919, another half-track intended for World War I was the Four Wheel Drive (FWD) Auto Company Model B truck converted to a balloon winch carrier by the McKeen Motor Car Company. This vehicle also replaced the rear wheels with a small Holt tractor suspension, but it retained the powered front wheels of the FWD truck.

The Four Wheel Drive Model B truck can be seen below converted to the half-track configuration. The vehicle at the left is equipped as a balloon winch carrier with a second engine at the rear to power the winch. Note the radiator at each end of the vehicle.

Two photographs of the Jeffrey/Nash Quad are shown above. The action of the rear mounted track suspension when climbing over an obstacle can be seen in the right view.

Another half-track conversion of a truck with four wheel drive was the Jeffrey (later Nash) Quad. It was based upon a 2 ton Model 4017 4x4 truck. Once again, the Holt tracked suspension replaced the rear wheels. All of these conversions exhibited improved cross-country mobility, but they were too slow to operate with truck convoys.

During the postwar years, a variety of half-track conversions were evaluated based upon just about every suitable truck or tractor available. J. Walter Christie produced several half-track conversions based upon various models of the Mack truck. Although greater speeds were obtained, none of these vehicles were reliable enough for troop use.

Above are two Christie conversions of the Mack Model AC truck to a half-track. The vehicle at the left, photographed on 15 May 1922, was fitted with tracks around the solid tires on the four rear wheels, all of which were chain driven. The front wheels were not powered. The Christie conversion at the right utilized a new tracked suspension on the rear with pneumatic tires on the front wheels. This vehicle was tested at Aberdeen Proving Ground during June 1924. Neither of the Christie conversions was successful.

The two photographs below show the Mack "Roadless" based upon the AB series Mack truck. Evaluated at Aberdeen during 1923, it was well liked, but considered somewhat underpowered with its four cylinder, 28 horsepower engine.

The 10 horsepower Citroen-Kegresse half-track is shown above at Aberdeen Proving Ground on 4 August 1925. At the right is the Class BBW belted six wheel five ton truck assembled by the Quartermaster Corps at Fort Holabird Maryland. The latter photograph was dated 1929.

An evaluation of the French Citroen-Kegresse half-track at Aberdeen Proving Ground provided a further advance in the development program. Two of these vehicles powered by 10 horsepower engines were purchased in 1925 for consideration as prime movers for the 75mm gun.

In 1931, a later version of the Citroen-Kegresse half-track was procured for test at Aberdeen. Designated as the P-17, it weighed 4300 pounds and was driven by an engine developing approximately 28 horsepower at 2500 rpm. It could pull 3500 pounds and carry a load of about 1000 pounds. Maximum speed was approximately 18 miles per hour. Front wheel steering was combined with a braking action on the 9 inch wide tracks. The overall dimensions of the P-17 were $167\frac{1}{2}$ inches long, $62\frac{1}{2}$ inches wide and $78\frac{1}{2}$ inches high. Although the test results were promising, both types of Citroen-Kegresse vehicles were considered to be too small and under powered for the proposed application. Neither of the Citroen-Kegresse vehicles had front wheel drive.

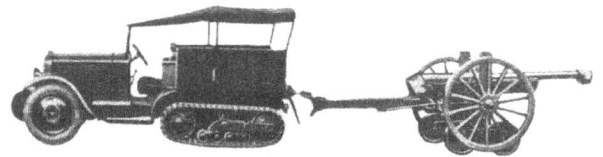

The photograph below and the two views above at the right show the Citroen-Kegresse P-17 at Aberdeen Proving Ground during June 1931. In the lower photograph above, the P-17 is being used in its role as a prime mover for the 75mm gun.

The half-track car T1 can be seen in the two photographs above. The original configuration of half-track car T1E1 appears at the bottom of the page during the tests in the Summer of 1933. Note the larger body and strengthened idler frame.

HALF-TRACK CARS

After the evaluation of the Citroen-Kegresse half-tracks, procurement was authorized for a pilot half-track from the James Cunningham, Son and Company of Rochester, New York. The new vehicle was designated as the half-track car T1 by Item 9957 of the Ordnance Committee Minutes (OCM) dated 7 July 1932.

The T1 had been built during 1931 and was larger and more powerful than the Citroen-Kegresse vehicles. It was fitted with a leaf spring bogie suspension and a new $6\frac{1}{2}$ inch wide track with a 2 inch pitch consisting of a roller chain with rubber blocks. The sprocket drive was at the front of the tracked suspension with a spring loaded idler at the rear. The front wheels, fitted with 7:00 x 20 tires, had a tread of $63\frac{3}{8}$ inches. They were not powered. The T1 had an empty weight of 6300 pounds and carried a load of 1900 pounds. It measured 178 inches in length, 81 inches in width, and 66 inches in height. The wheelbase was 105 inches. The vehicle was powered by a Cadillac V8 engine developing 115 horsepower at 3200 rpm. It had a maximum road speed of 42 miles per hour with five speeds forward and one reverse.

After testing, some modifications were made. These included a strengthened idler frame and a new body with a larger passenger compartment and driver's seat.

Thirty of the modified vehicles, now designated as the half-track car T1E1, were assembled at Rock Island Arsenal from components procured from James Cunningham, Son and Company. The specifications were much the same as the T1 except the weight with the new body had increased to 6620 pounds and the cargo capacity was now 2000 pounds. The overall dimensions increased to $180\frac{1}{2}$ inches in length, $82\frac{1}{2}$ inches in width and 77 inches in height. These vehicles were issued to the Cavalry. In March 1939, OCM 14965 belatedly recommended that they be classified Limited Standard as the half-track car M1. During their troop service, the vehicles were used in the development of tactics and various modifications such as the front mounted roller for use in crossing ditches. In December 1940, they were declared obsolete by order of the Secretary of War.

The original half-track car T1 pilot was used to evaluate a number of modifications. The spring loaded idler was replaced by a rigidly mounted, adjustable, unit and the new T6 rubber jointed track was installed. Unfortunately, this track failed during the tests at Aberdeen Proving Ground which ran from June to October 1933. Ordnance Committee action designated this modified vehicle as the half-track car T1E2.

Additional modification of the original pilot replaced the leaf spring track suspension with a volute spring articulating bogie. Now designated as the half-track car T1E3, the pilot was evaluated at Aberdeen during 1934. This new volute spring track suspension was the forerunner of those adopted for the half-tracks during World War II.

A half-track car T2 was the subject of a design study, but it was never built.

Interior details of the body on the half-track car T1E1 are visible in the view above at the left. A rear door is now provided and the spare tire is stowed at the front of the troop compartment. The T1E1 as modified for production appears at the top right. The suspension has been redesigned with a fixed idler replacing the spring loaded version. The vehicle is operating on the washboard course at Aberdeen Proving Ground on 22 August 1934.

The half-track car T1E2 is above at the left. It is still marked as the half-track car T1. Details of the fixed rear idler and the track can be seen above at the right. The reworked half-track car T1 is shown again below. Now it is designated as the half-track car T1E3 with the volute spring bogie replacing the earlier leaf spring suspension. The original short body is obvious and it still carries the marking half-track car T1. The photograph at the left was taken at Aberdeen Proving Ground on 25 September 1934. The right photograph was dated 28 August 1934 at Rock Island Arsenal.

Some of the thirty half-track cars M1 (T1E1) are shown here in service with the troops. They are armed with .30 caliber M1919A4 machine guns and the two vehicles immediately below are equipped with radios.

The half-track truck T1 can be seen above at the left at Aberdeen Proving Ground on 24 August 1933. At the top right, the T1 is towing a 75mm howitzer at Aberdeen on 7 January 1936.

HALF-TRACK TRUCKS

In 1933, parallel with the work on the half-track cars, development was authorized for the half-track truck T1. This vehicle was converted from a standard General Motors $2\frac{1}{2}$ ton Model T-33 truck. A tracked suspension, similar to that on the half-track car T1E2, replaced the rear axle and wheels. A two speed auxiliary transmission was installed in addition to the standard four speed transmission providing a total of eight speeds forward and two in reverse. Fitted with a new cab and body, the T1 was tested at Aberdeen from August to November 1933. The six cylinder GMC truck engine developed 96 horsepower at 3000 rpm providing a maximum speed of 41 miles per hour. The T14 rubber block track was $6\frac{1}{2}$ inches wide with a 2 inch pitch. The T1 half-track truck weighed 7780 pounds and carried a load of 2000 pounds. Its overall dimensions were 197 inches in length, 78 inches in width, and 85 inches in height. The tests at Aberdeen showed superior performance when the T10 rubber band track was installed. However, its durability was unsatisfactory and further tests were suspended until a satisfactory band type track could be developed. As a result, the six vehicles then on order were canceled.

Two views of the Cunningham-Chevrolet half-track truck are at the right. The upper photograph was dated 8 August 1934. Below is a photograph of the Cunningham-Ford half-track truck at Aberdeen Proving Ground on 10 August 1933.

A proposed modification of the half-track truck T1 was the half-track scout car T6. It was intended to have two "swivel chair" type turrets. However, it was not built.

Another vehicle evaluated at Aberdeen from August to December 1933 was the half-track truck T2. This vehicle was a conversion of a standard 1933 Ford $1\frac{1}{2}$ ton truck by the installation of a Cunningham rear track suspension similar to that on the half-track car T1E1. It proved to be very satisfactory during the tests, but it was barred from procurement because of the manufacturer's non-compliance with the NRA codes then in effect.

Another conversion, similar to the half-track truck T2, was made using the Chevrolet 1 ton truck. It also was tested at Aberdeen during August 1934 fitted with the Cunningham T17 rubber block track. However, it was inferior to the T2 Ford conversion because of its lower power engine.

The Linn half-track truck T3 is shown above. At the right, the Linn half-track is carrying the light tank T1E6.

The half-track truck T3 was a higher powered version of the standard Linn Tractor Model WD-12. The new engine was the American LaFrance V12 developing 222 horsepower at 2400 rpm. The T3 weighed 18,230 pounds and had a normal cargo load of 16,000 pounds with a maximum load of 20,000 pounds. The T3 was 268 inches long, 93 inches wide, and 78 inches high. Its front wheels were not powered. Fitted with 9:75 x 20 tires, the front tread was $55\frac{1}{2}$ inches. The Linn track was 14 inches wide with a pitch of 8 inches. A top speed of 19.6 miles per hour could be reached on level roads.

The half-track truck T4 was similar to the half-track truck T1, but it was a new design with the objective of increasing the performance and life. It was a commercial General Motors truck with the rear wheels replaced by the Cunningham tracked suspension. The T4 weighed 8100 pounds and its overall dimensions were 209 inches in length, 81 inches in width, and $98\frac{1}{2}$ inches in height. The front wheels were not powered. They had a tread of

$60\frac{1}{4}$ inches and were fitted with 7:00 x 20 tires. The T4 weighed 8100 pounds and carried a cargo of 2000 pounds. The six cylinder GMC engine developed 90 horsepower at 2500 rpm. With a four speed standard transmission plus a two speed auxiliary transmission, the vehicle had eight speeds forward and two in reverse. The maximum road speed was about 26 miles per hour.

The half-track truck T4E1 was a specification intended for the procurement of wire laying trucks for the Signal Corps. None of these were produced since wheeled vehicles were considered preferable for this application.

The wire laying half track truck T4 appears in the photographs below fitted with the T10 rubber band track. The vehicle was being evaluated at Aberdeen Proving Ground on 15 June 1934.

A half-track truck T5 is shown above without the rear canvas cover. Below the rear cover as well as the side curtains have been installed.

Tests of the half-track truck T1 indicated that its higher speeds were desirable for a prime mover of light artillery. To meet this requirement, General Motors Truck Corporation produced the half-track truck T5. It utilized many commercially available components and the rear wheels were replaced by the Cunningham articulating bogie, leaf spring, tracked suspension. The T14 track was $8^{7}/_{8}$ inches wide with a 2 inch pitch. Goodrich sponge filler tires were installed on the front wheels with a tread of 63 inches. The T5 was 222 inches long, 80 inches wide, and 96 inches high. It weighed 8840 pounds empty and 12,580 pounds with a load. The six cylinder engine produced 125 horsepower at 2800 rpm providing a maximum road speed of 39.2 miles per hour. A total of 24 T5s were manufactured.

During the test program, a T5 was converted to the T5E1 by the installation of a differential with a ratio of 7.45:1 replacing the differential with a ratio of 5.57:1. This change reduced the maximum speed to 29.3 miles per hour. Tests at Aberdeen from October 1935 to September 1936 indicated that the T5E1 was inferior to the T5 when towing the 155mm howitzer.

The original T5 was fitted with 10 inch wide bogie wheels and tracks and designated as the half-track truck T5E2. Tests at Aberdeen showed the wider tracks and bogies wheels to be far superior to the narrow type.

The wide track suspension of the half-track truck T5E2 appears above. Below are views of another half-track truck T5.

The production version of the half-track truck T5 can be seen in these photographs in service with the troops. Above, the 75mm howitzer M1 is in the firing position. Below, it is being towed by the half-track truck.

Below are additional views of the half-track truck T5 in its role as the prime mover for the 75mm howitzer M1. The normal stowage can be seen on these vehicles.

The Linn half-track truck T6 appears in the photographs above and below. The top photograph shows the vehicle as furnished to the U.S. Army in Hawaii with the rear canvas cover and the side curtains installed. In the bottom view, the T6 is under test at Aberdeen Proving Ground on 13 July 1934.

The half-track truck T6 was another product of the Linn Manufacturing Company. It was similar to the T3 described earlier, but somewhat smaller and a little heavier. It was rated for the same 16,000 pound cargo load as the T3. The overall dimensions were 246 inches long, 92 inches wide, and 98 inches high. The Linn 14 inch wide, 8 inch pitch, manganese steel track was used as on the T3. A Hercules HXE engine developed 174 horsepower at 1600 rpm. In addition to a main transmission with five speeds forward and one reverse, there was a reverse transmission which allowed all of the main transmission speeds to be used in the reverse direction. The maximum road speed of the vehicle was 15.7 miles per hour. It was tested first at Fort Bragg and later at Aberdeen Proving Ground demonstrating excellent reliability.

The half-track truck T7 was a designation assigned for a vehicle suitable for use as a prime mover for medium artillery. No vehicles were produced under this designation.

The half-track truck T8 was a modified Ford $1\frac{1}{2}$ ton truck chassis with an extended frame and the Trackson tandem drive attachment. It was fitted with the Trackson cable track around the two rear tires. This track was $15\frac{3}{4}$ inches wide providing a low ground pressure. Tests at Aberdeen revealed the system to be unreliable and no further development was recommended.

At the right is the half-track truck T8 with the Trackson tandem drive attachment and cable tracks on 6 February 1937.

The production version of the half-track truck T9 can be seen above and below in these photographs at Aberdeen Proving Ground dated 28 January 1938. This was the vehicle subsequently standardized as the half-track truck M2.

The half-track trucks T9 and T9E1 were identical except for the tracked suspension. Both were based upon a 1936 Ford truck chassis modified with a Marmon-Herrington front wheel drive. It was fitted with a pressed steel cab and a 1½ ton capacity cargo body. The 1936 Ford V8 engine developed 83 horsepower at 3800 rpm. The standard transmission was modified and assembled to a Marmon-Herrington auxiliary transmission with an over-running clutch in the gearing to the front wheels. This provided eight forward and two reverse speeds. On the initial installation, the front wheels could not be driven in reverse. The volute spring bogie on the T9 had four 12 inch diameter steel wheels on each side. On the half-track truck T9E1, the bogie had two 20 inch diameter wheels on each side. These could be either steel or solid rubber tired wheels. The T9 was fitted with a T21 rubber block track 10 inches wide with a 5 inch pitch. The T9E1 used a T20E2 rubber block track 8 inches wide with a 5 inch pitch. The empty weights of the T9 and the T9E1 were 8560 pounds and 8410 pounds respectively. The cargo load was 3250 pounds in both cases. The maximum speeds were 25 miles per hour for the T9 and 35 miles per hour for the T9E1.

At the right is a close-up view of the suspension for the half-track truck T9 with the bogie wheel jammed during a run on the washboard course.

The production version of the half-track truck T9E1 was photographed at Aberdeen along with the T9 on 28 January 1938

Two production models each were built of the T9 and T9E1 with the following changes. The front wheel drive was modified so that it would operate in reverse, two hand brakes were added on the sprocket shafts to aid in steering, the battery was shifted from the right running board to the upper right on the engine side of the dash, a new military body with seats for six was installed, and the suspension on both vehicles was modified to use the 10 inch wide T24E1 rubber band tracks. For a time, the T9 was standardized as the half-track truck M2.

The pilot T9E1 was fitted with special pneumatic tires on the bogie wheels and designated as the half-track truck T9E2. Tests at Aberdeen had unsatisfactory results and the vehicle was returned to its original state.

Tests of the T9 series vehicles proved the feasibility of synchronizing the front wheel and track drives on the half-track vehicles.

The half-track truck T10 was a proposed light half-track truck intended for wire laying. It was never built.

At the left is a close-up view of the pneumatic tired bogie wheels with and without the side flange as installed on the half-track truck T9E2. This photograph was dated 30 July 1937.

Above is another view of the production half-track truck T9E1. Below, the earlier T9E1 is at the left on 25 January 1937 and the earlier T9 is at the right on 29 October 1937.

Below, the half-track truck M2(T9) is in service with the troops towing the 75mm howitzer M1 over rough terrain.

PART II

HALF-TRACKS FOR WORLD WAR II

Above, the half-track personnel carrier T7 shows its close relationship to the scout car M2A1 from which it was converted.

ARMORED HALF-TRACKS

The first armored half-track resulted from a project authorized by OCM 14188 to convert an M2A1 scout car to a half-track vehicle. Some sources refer to the vehicle converted as the M3 scout car. Since the designation of the scout car M2A1 was subsequently changed to M3, both sources are correct. The work was carried out at Rock Island Arsenal with help from the White Motor Company. Subsequently designated as the half-track personnel carrier T7, there was no change in the chassis, the 95 horsepower Hercules engine, or the transmission. The converted vehicle retained the front wheel drive of the scout car and two tracked, volute spring, suspensions with rubber band tracks similar to those on the half-track truck T9 replaced the rear wheels. Larger front tires were installed as well as shorter front and longer rear propeller shafts. A transfer case with a de-clutching device was added. The gear ratio was changed in the front and rear axles. The face hardened armor plate was $1/4$ inch thick except on the windshield cover where it was $3/8$ inches thick. The total weight of the T7 was 12,170 pounds with a payload of 1800 pounds. It was intended to carry a crew of eight men with two .30 caliber machine guns and one .50 caliber machine gun.

Tests at Aberdeen Proving Ground extended from 21 September to 24 October 1938. It was noted that the front wheel drive gave the T7 superior cross-country performance, but the vehicle was under powered. A front mounted roller was installed during the tests and proved to be of great value in crossing ditches. After completion of the test program, the T7 was converted back to a scout car and returned to the troops.

Details of the half-track personnel carrier T7 can be seen in the photographs below. The vehicle is shown with and without its canvas top.

The half-track scout car T14 appears above and in the left photograph at the bottom of the page. It is now equipped with the front mounted roller evaluated on the half-track personnel carrier T7.

By 1939, it was becoming obvious that the Army would require large numbers of half-track vehicles during its future expansion. As a result, on 26 December, the Artillery Division, Industrial Service of the Ordnance Department submitted drawing D-42876 and a specification for an armored half-track incorporating the results of the various experimental programs. They recommended the construction of a pilot vehicle. The Ordnance Committee approved two days later in OCM 15544, dated 28 December 1939, designating the new pilot as the half-track scout car T14. It was to be the basis for all of the American half-tracks to see action during World War II.

The pilot T14 was built during the first part of 1940 at the White Motor Company in Cleveland, Ohio. Its first operation was an overland road march from Cleveland to Aberdeen Proving Ground, Maryland on 28-29 May 1940. The test program at Aberdeen continued until 28 September 1940.

As delivered, the T14 did not completely meet the specification requirements. To expedite construction of

the pilot, the tracked suspension from the half-track personnel carrier T7 was installed on the T14. This suspension had the drive sprocket at the rear and was fitted with steel bogies and 10 inch wide tracks. The specification required front mounted sprockets and rubber tired bogies with 12 inch wide tracks. Also, an engine with a maximum torque rating of 325 foot-pounds was specified. The original engine installed in the pilot was the White Model 20A which produced a maximum torque of 280 foot-pounds at 1200 rpm. This was a six cylinder L-head engine developing 116 horsepower at 3000 rpm. However, four additional six cylinder engines were evaluated in the T14 during the test program. These were the White 140A, the White 160A, the Hercules WXLC3, and the Buick Series 60. The latter was a valve in head passenger car engine which produced 142 horsepower at 3600 rpm. The highest performance was obtained with the Buick engine and the White 160A. The latter engine developed 147 horsepower at 3000 rpm and had excellent torque characteristics.

The hood of the T14 at the right is modified for the Buick engine.

Note the rubber tired bogie wheels on the half-track scout car T14 above in these photographs dated 4 December 1940. Compare these with the original bogie wheels in the close-up view of the suspension below.

Even before the Proving Ground test reports were published, the Ordnance Committee acted to standardize three basic half-track vehicles. OCM 16112, dated 19 September 1940, recommended that the modified T14 be standardized as the half-track car M2 with seats for 10 men. With the body and frame of the vehicle extended 10 inches to the rear, it was standardized as the half-track personnel carrier M3 with a capacity of 13 men. This version was developed by the Diamond T Motor Car Company as the half-track personnel carrier T8. When the T14 body was modified to carry an 81mm mortar with its crew and ammunition, it was standardized as the 81mm mortar carrier M4. All three chassis had the same 135.5 inch wheelbase as measured from the center of the front wheels to the center of the track assembly.

This photograph of the T14, also dated 4 December 1940, clearly shows the new bogie wheels and the armament consisting of one .50 caliber M2HB machine gun and one .30 caliber M1919A4 machine gun. Note that the side compartment door is hinged along the side so that it swings open to the rear.

These two views show the half-track scout car T14 during its evaluation at Aberdeen Proving Ground on 10 December 1940. In the photograph above, it is towing the 105mm howitzer. The rail for the skate mounted machine guns is clearly visible.

Two weeks after standardization was recommended, the White 160A engine was selected for all of the production half-tracks. The cooling system was redesigned to accommodate the higher power engine and it was provided with an adequate oil bath air cleaner. The tests of the T14 had shown excessive track wear. However, it was expected that this would be reduced with the production suspension using rubber tired bogie wheels and the 12 inch wide track. The front mounted sprockets also were expected to reduce the damage to the track guides.

By September 1940, the requirements for half-track vehicles had increased beyond the capacity of any single manufacturer and orders were placed with the Autocar Company, the Diamond T Motor Car Company, and the White Motor Company. On 28 September and 1 October, meetings were held in Washington D. C. and Ardmore, Pennsylvania resulting in the formation of the Half-Track Engineering Committee. Consisting of representatives of all three companies and the Ordnance Department, it was responsible for the design of the half-tracks and to ensure that all parts, except armor plate, were interchangeable between the vehicles built by the three manufacturers.

By mid October 1940, the standardization of the three vehicles had been approved and procurement was authorized.

A privately developed armored half-track also appeared during 1940. This was the Marmon-Herrington DHT-5. Similar in appearance to some of the Marmon-Herrington prewar armored cars, it was equipped with a turret mounted 37mm gun as main armament. The rear tracked suspension appeared to be a heavier version of that installed on the half-track trucks T9 and T9E1. The DHT-5 was not evaluated by the U. S. Army.

Below is the Marmon-Herrington DHT-5 half-track armed with the turret mounted 37mm gun. Note the vertical volute suspension in the half-track bogie.

Above, the first production half-track car M2 is accepted by the U.S. Army at the White Motor Company.

HALF-TRACK CARS AND PERSONNEL CARRIERS

The first production vehicles were accepted in May 1941 with 62 M2 half-track cars delivered by the White Motor Company. This began the long run that would produce 11,415 new M2s from Autocar and the White Motor Company until it was succeeded on the production lines by the later M2A1.

As mentioned earlier, the half-track car M2 was similar to the half-track scout car T14 which served as a pilot for all of the half-track vehicles. Like the T14, the M2 was protected by face hardened steel armor $1/4$ inch thick on all surfaces except for the $1/2$ inch thick plate over the windshield. The open top vehicle was provided with a canvas top for bad weather, but it was rarely used in a combat area because of the restricted visibility and interference with the skate mounted armament.

Initially, the armament specified for the M2 consisted of one air-cooled, heavy barrel, .50 caliber machine gun M2 and two water-cooled .30 caliber machine guns M1917A1. The water-cooled weapons were replaced later by a single air-cooled .30 caliber machine gun M1919A4. All of these weapons were on skate mounts riding on the gun rail that surrounded the open top of the vehicle just below the top of the armor.

At the right, M2 half-track cars are lined up for delivery at the White Motor Company.

29

The half-track car M2 above is at the Raritan Arsenal on 23 June 1941. Note that the stowage compartment side doors in the hull are now hinged at the bottom. Below is a view of the M2 chassis with all of the armor plate removed.

The M2 was intended as an armored combat and reconnaissance vehicle and as a prime mover for artillery. Two large ammunition storage compartments were installed, one on each side of the vehicle. The top of these could be opened providing access to the top shelf from inside the vehicle. The lower shelves could be reached from the outside through doors in the side armor. These side doors were hinged at the bottom, unlike those on the T14 which were hinged at the side.

The ten seats in the M2 were arranged with two in the driving compartment and two between the two storage compartments. The latter were back to back facing to the front and to the rear. Three additional seats were on each side in the back facing toward the center. A 30 gallon self sealing fuel tank was installed on each side at the rear of the crew compartment.

The pilot half-track car M2 appears below at Aberdeen Proving Ground on 9 April 1941. The armor flaps on the side doors and the windshield cover are closed and the vehicle is armed with one .50 caliber M2HB machine gun and two .30 caliber M1917A1 machine guns.

Above is a later production half-track car M2 fully stowed with the stowage compartment open. This vehicle is fitted with mine racks and is armed with one .50 caliber M2HB machine gun and two .30 caliber M1917A1 machine guns. Below are two views of an early M2 with the canvas cover installed.

Further details of the early half-track car M2 can be seen below. Note the early headlights on this vehicle. The rail for the skate mounted machine guns is visible in the left photograph.

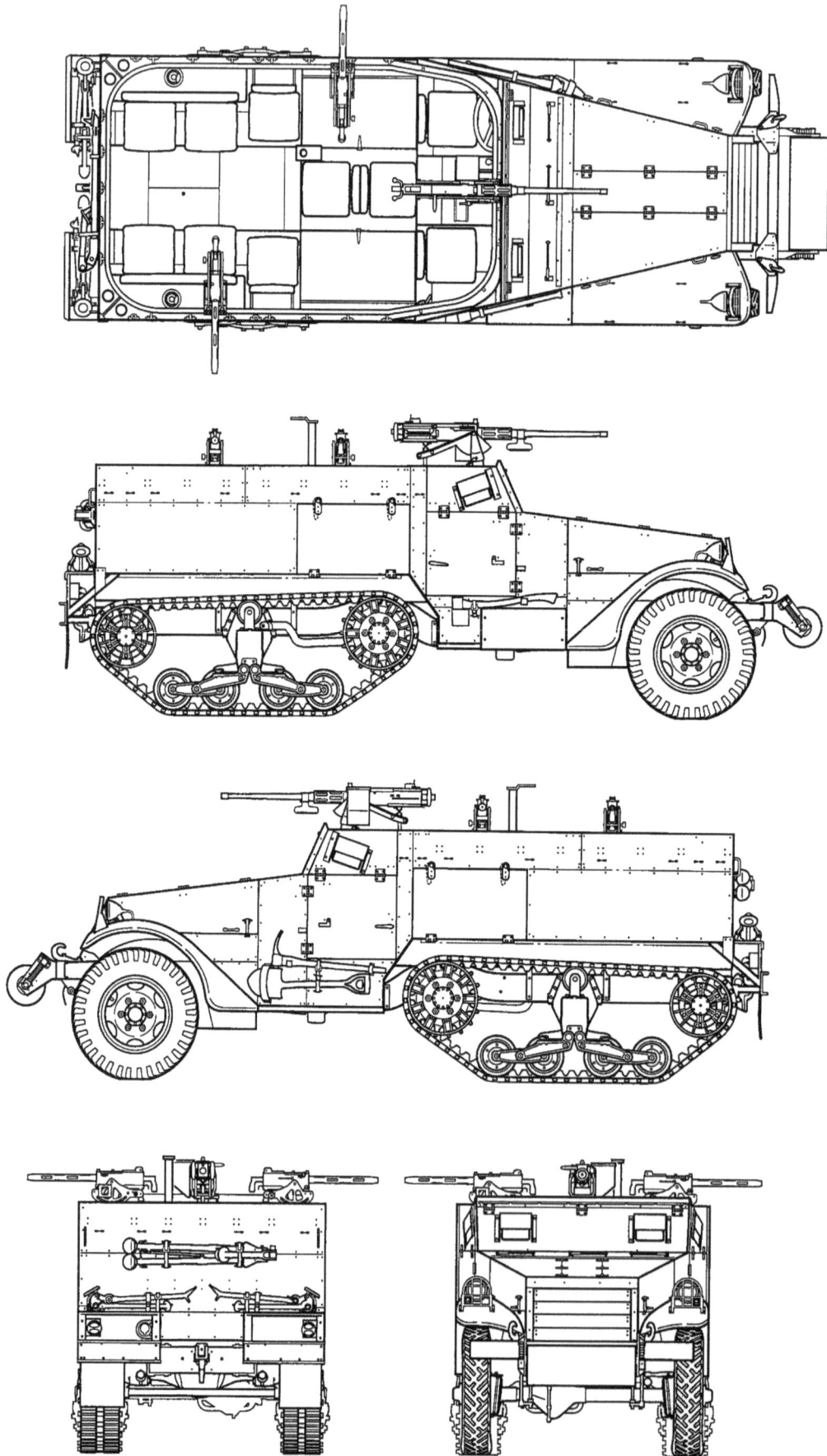

Scale 1:48

©M. Duplessis

Half-Track Car M2, early production

The early half-track personnel carrier M3 above and below is at Aberdeen Proving Ground on 27 June 1941. The vehicle is armed with a single .30 caliber M1919A4 machine gun on a pedestal mount.

A single M3 half-track personnel carrier also was received in May 1941 from the Diamond T Motor Car Company. Production of new M3s would increase from that single vehicle to 12,391 by the time production shifted to the later M3A1. The M3 was produced by all three manufacturers.

The M3 half-track was developed as a personnel carrier for armored infantry, but it proved to be extremely adaptable for many applications. Among these were an ambulance, a command vehicle, an engineer vehicle, a radio carrier, a prime mover for artillery, as well as a chassis for several expedient self-propelled weapons.

The armor protection on the M3 was the same as on the M2, but, as mentioned before, the body was ten inches longer and it was fitted with a door in the rear to permit easy passage in and out of the vehicle. The storage compartments on the M2 were eliminated and the two 30 gallon self sealing fuel tanks were shifted forward to just behind the driving compartment. Three seats were in the driving compartment with the center seat slightly to the rear. Five additional seats were installed on each side of the crew compartment bringing the total to thirteen. A .30 caliber M1919A4 machine gun was installed on a pedestal mount in the crew compartment. Frequently, the troops would replace the .30 caliber weapon with .50 caliber machine gun.

The early half-track personnel carrier M3 at the right is fitted with the canvas top.

33

The photographs on this and the following page show the early half-track personnel carrier M3. In the views above and at the left, the side armor flaps on the doors are closed, the windshield cover is open, and the canvas top has not been installed. At the bottom of the page, the armor door flaps are open and the canvas top and side curtains are in place.

These photographs of the early half-track personnel carrier M3 show many details of the vehicle. The early type headlights and the front roller are clearly visible above. Note the directional tire tread on this early vehicle. The interior arrangement and the 13 crew seats can be seen below and at the right.

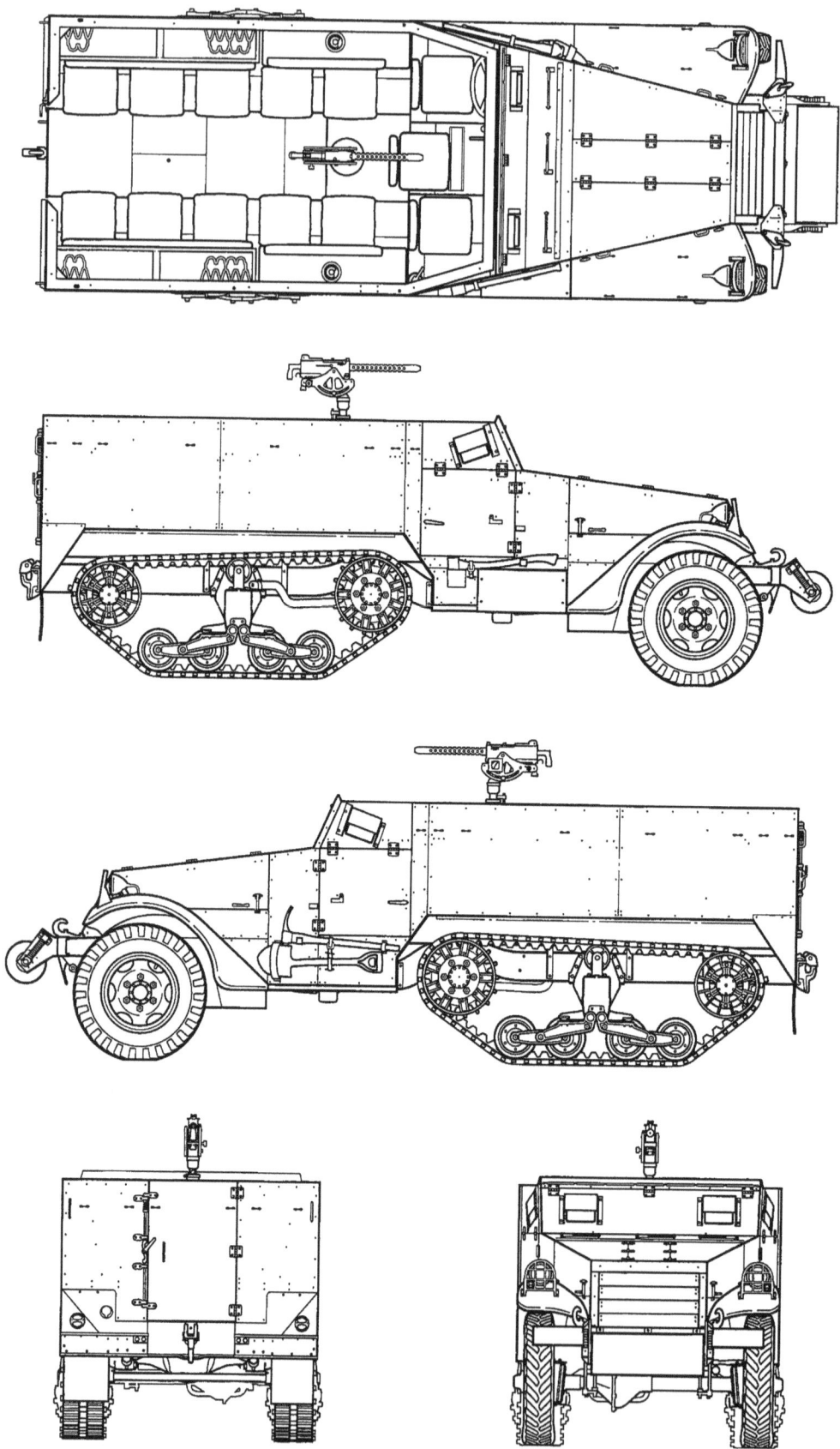

Scale 1:48

Half-Track Personnel Carrier M3, early production

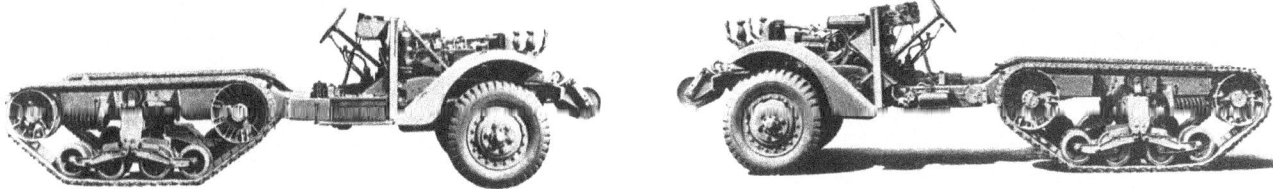

The late production chassis for the half-track car M2 can be seen above and below at the right. Note the double coil, spring loaded, idler and the late type headlights. The frame of the half-track is in the drawing below. The letters in the drawing indicate the dimensions to be taken when checking the alignment of the frame.

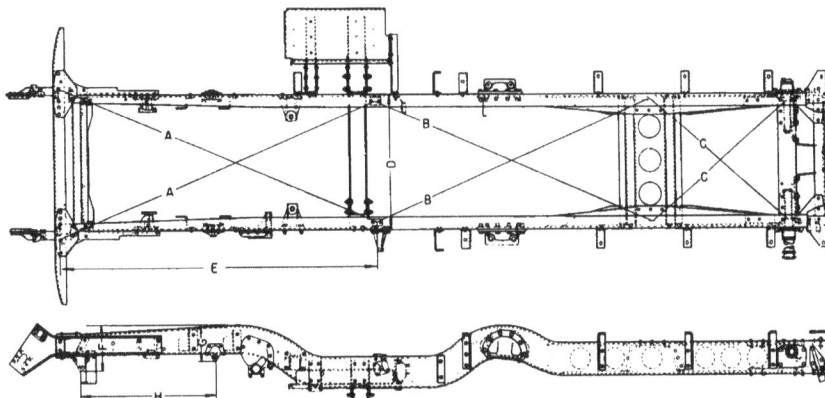

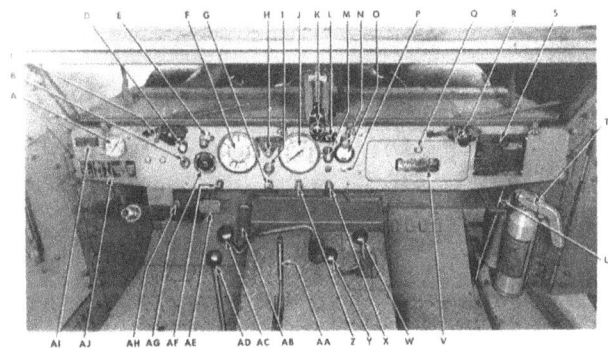

Above are the driver's instruments and controls on the half-tracks manufactured by White, Autocar, and Diamond T. They are identified as follows:

A. tachometer, B. trouble light receptacle, C. left wiper control, D. panel light control, E. main light switch, F. instrument cluster, G. throttle, H. starter button, I. Ignition, J. speedometer, K. compass, L. fuel tank selector, M. blackout light, N. dash light, O. voltmeter, P. voltmeter button, Q. map compartment, R. right wiper control, S. registration plate, T. fire extinguisher, U. radiator shutter control, V. winch caution plate, W. front drive shift lever, X. right ventilator control, Y. transfer case shift lever, Z. choke, AA. parking brake, AB. accelerator, AC. transmission shift lever, AD. power take off shift lever, AE. brake, AF. left ventilator control, AG. electric brake load control, AH. clutch, AJ. gear shift plate, AI. engine caution plate.

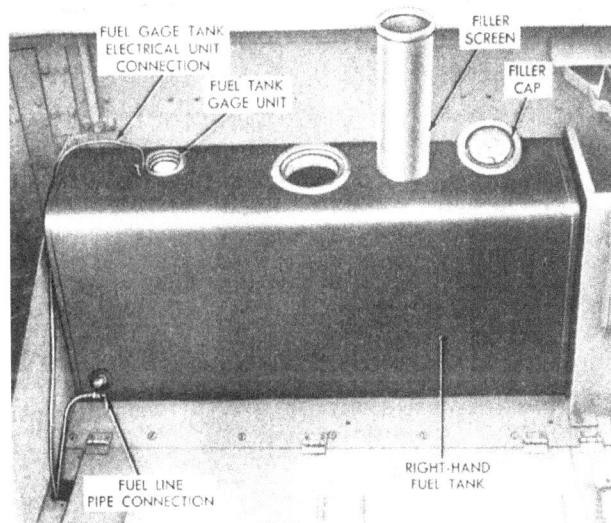

Details of the right-hand fuel tank can be seen above. The particular installation shown is in the half-track personnel carrier M3. Note the rifle rack at the right side of the photograph. At the left is a sketch of the fuel system on the half-tracks except for the M15 and M15A1 motor carriages. The latter had one vertical fuel tank to the rear of the driving compartment and one horizontal fuel tank under the front section of the turret.

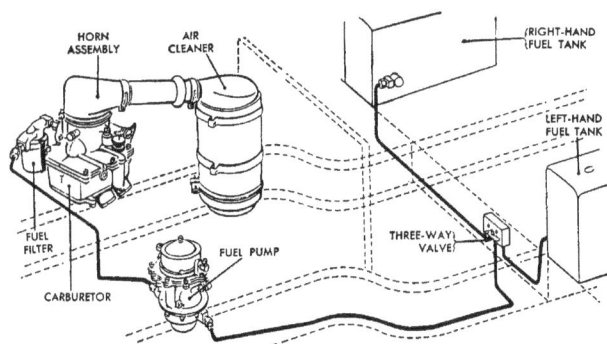

37

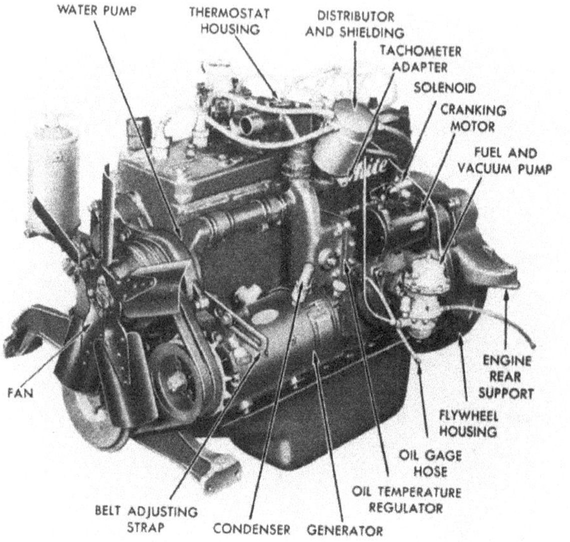
WATER PUMP
THERMOSTAT HOUSING
DISTRIBUTOR AND SHIELDING
TACHOMETER ADAPTER
SOLENOID
CRANKING MOTOR
FUEL AND VACUUM PUMP
FAN
ENGINE REAR SUPPORT
FLYWHEEL HOUSING
OIL GAGE HOSE
OIL TEMPERATURE REGULATOR
BELT ADJUSTING STRAP
CONDENSER
GENERATOR

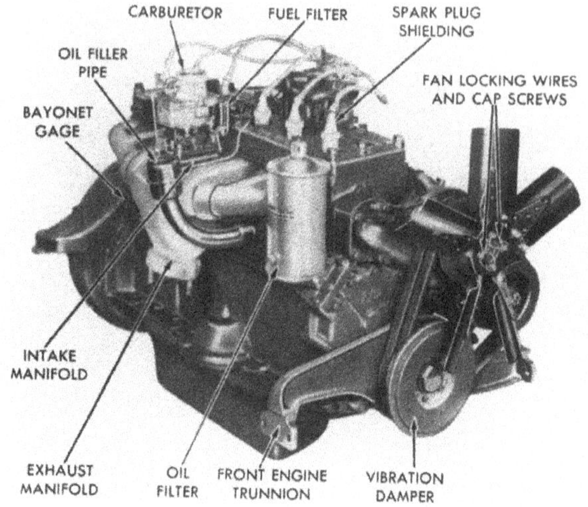
CARBURETOR
FUEL FILTER
SPARK PLUG SHIELDING
OIL FILLER PIPE
BAYONET GAGE
FAN LOCKING WIRES AND CAP SCREWS
INTAKE MANIFOLD
EXHAUST MANIFOLD
OIL FILTER
FRONT ENGINE TRUNNION
VIBRATION DAMPER

The White 160AX engine is shown above with a cutaway view at the right. Close-up views of the front roller and the winch appear at the bottom of the page.

The power train in the production half-tracks consisted of the White 160AX gasoline engine driving the vehicle through the Spicer 3461 transmission. This transmission had four speeds forward and one reverse and it engaged and disengaged the front wheel drive. A two speed transfer case was bolted to the rear of the transmission. Its high and low ranges doubled the number of speeds available. The front propeller shaft transmitted power to the front axle and then to the wheels. The front wheels were fitted with 8.25 x 20 tires. Early vehicles used self sealing tires, but later half-tracks were fitted with 12 ply combat tires on split bead rims with metal bead locks. The front wheels supported the front of the vehicle through leaf springs.

The early production half-tracks were fitted with the front mounted roller to aid in crossing ditches and other rough terrain. The later vehicles had either the roller or a 10,000 pound capacity Tulsa Model 18G winch mounted just behind the front bumper. A power take-off on the transmission supplied power to the winch.

ROLLER

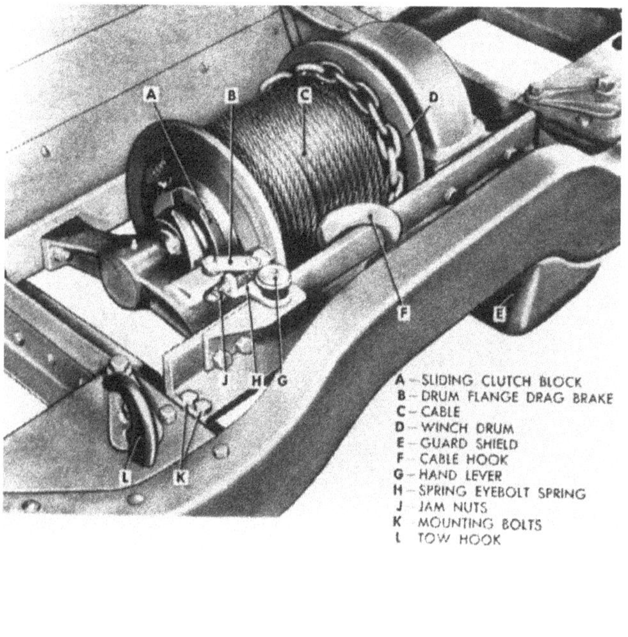
A
B
C
D
E
F
G
H
J
K
L
A – SLIDING CLUTCH BLOCK
B – DRUM FLANGE DRAG BRAKE
C – CABLE
D – WINCH DRUM
E – GUARD SHIELD
F – CABLE HOOK
G – HAND LEVER
H – SPRING EYEBOLT SPRING
J – JAM NUTS
K – MOUNTING BOLTS
L – TOW HOOK

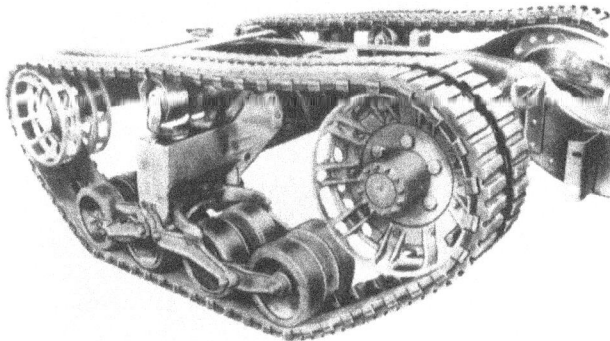

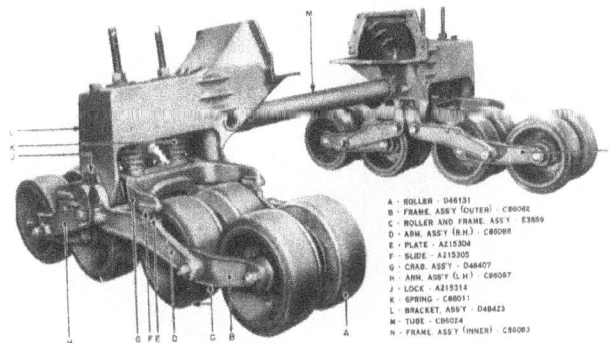

A - ROLLER - D46131
B - FRAME, ASS'Y (OUTER) - C86082
C - ROLLER AND FRAME, ASS'Y - E3859
D - ARM, ASS'Y (R.H.) - C86088
E - PLATE - A215304
F - SLIDE - A215305
G - CRAB, ASS'Y - D46407
H - ARM, ASS'Y (L.H.) - C86087
J - LOCK - A215314
K - SPRING - C86011
L - BRACKET, ASS'Y - D48423
M - TUBE - C86024
N - FRAME ASS'Y (INNER) - C86083

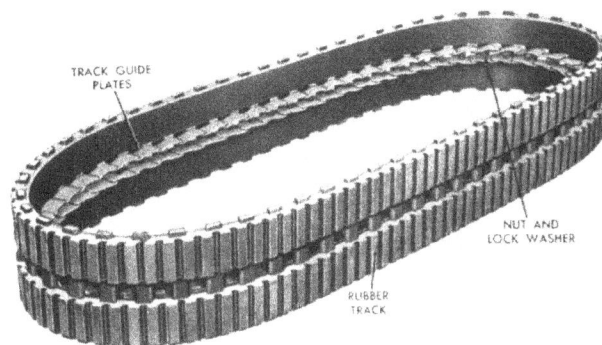

Above, the early half-track suspension with the fixed idler is at the left and details of the bogie can be seen at the right. The Goodrich rubber band track assembly is at the right.

The rear propeller shaft powered the rear axle and then the track drive sprockets. These 18 tooth sprockets at the front of the rear suspension drove the rubber band tracks developed by Goodrich. This track consisted of rubber molded around steel cables bolted together by steel cross pieces to which center guides were attached. The cross pieces extended out of the track edge to aid in traction. Track tension was set by the adjustable fixed idler wheel at the rear of the tracks. The major part of the vehicle weight was carried on the bogie fitted with two vertical volute springs. The springs transmitted the load through the crab to the suspension arms and then to the four rubber tired rollers on each side of the vehicle. A steel roller on top of the track frame assembly supported the upper track run.

Hydraulic brakes were installed on the front wheels and the track drive sprockets. A mechanical parking brake was fitted on the rear propeller shaft.

On 21 August 1942, the Office, Chief of Ordnance directed that mine racks be added to each side of the vehicles. On the M2, these were fitted to the rear of the doors to the ammunition storage compartments. On the M3 they extended the full length of the sides. Modification Work Order G102-W21, dated 24 February 1943, was issued for the installation of mine racks on half-tracks in the field.

Details of the pintle hook and the bumperette on the half-track personnel carrier M3 are shown below at the left. At the bottom right, this half-track car M2 at Fort Knox has been fitted with mine racks. Note the early type headlights.

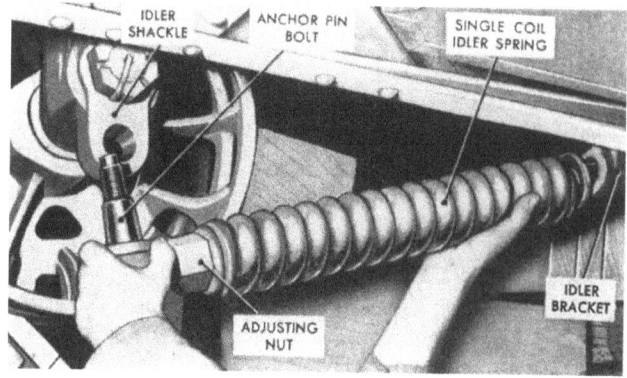

The early single coil, spring loaded, idler is illustrated in these photographs. At the right, the spring is being removed.

With the half-tracks in the hands of the troops, reports from the field as well as additional proving ground test results revealed the need for a number of modifications. A particularly serious problem involved the tracked suspension. As described earlier, this suspension had a front mounted sprocket and a fixed, but adjustable, rear idler to maintain track tension. It worked well for traveling on roads, but it could not compensate for track movement when crossing rough terrain. This often resulted in a thrown track or damaged suspension components. Ordnance was considering design modifications, but the 2nd Armored Division took matters into their own hands. Scheduled to leave for the invasion of North Africa in the Fall of 1942, their solution to the problem is described in the following excerpt from their history, "Hell on Wheels" by Donald E. Houston.

"The question of half-tracks came up and for a time it appeared that the infantrymen were on the verge of losing their personnel carriers, because the rear idler spindle was fixed in place and could not bend or give when moving over rough terrain. First Lieutenant Thomas Hauss and Master Sergeant Gerry Noble came up with a scheme to replace the fixed idler with an eyebolt and nut, and a coil spring from a Caterpillar tractor. Colonel Sidney R. Hinds personally paid for the items and directed that it be tested and that he be informed of the results. It was successful. Hinds took the idea to the division ordnance officer, Lieutenant Colonel Frederick Crabb, who took the suggestion to Major General Harmon. Harmon quickly approved it and had ordnance buy the modification parts and install the device on all the half-tracks slated for North Africa."

The Office, Chief of Ordnance had issued a directive on 4 September 1942 to provide a spring loaded idler for the half-track suspension and on 29 November 1942, Modification Work Order G102-W14 was issued covering a field fix to serve as an expedient until a production type spring loaded idler could be designed. This expedient was a single coil spring arrangement like that installed by the 2nd Armored Division, which by this time was in North Africa. On 15 July 1943, Modification Work Order G102-W36 was issued for the field installation of the new double coil idler spring then in production.

Operations with the half-track self-propelled artillery had shown that the headlights were frequently broken by the muzzle blast from the cannon. To eliminate this problem, a directive was issued to standardize demountable headlights for all new half-track vehicles. For vehicles already in the field, Modification Work Order G102-W34, dated 21 June 1943, was issued to modify only the artillery motor carriages.

The larger diameter older lights were fixed on the front fenders behind brush guards. A small blackout marker light was installed outboard of each headlight behind the brush guard. The new demountable lights were smaller and were located in brackets on each side of the hood armor. A combination headlight and black-out marker light was mounted behind a brush guard in each bracket. A combination blackout driving light and a blackout marker light could replace the headlight and blackout marker light only in the left bracket. The tail-lights also were modified on the later half-tracks.

Operations in the field, as well as the production of vehicles carrying heavier loads, resulted in increased weight on the various half-tracks. This caused a rapid increase in failures of the front springs. To correct this problem, the manufacturers were directed to install heavier front springs in the new production vehicles. This modification also was applied to the vehicles in the field.

A—IDLER WHEEL
B—IDLER SPRING
C—VOLUTE SPRING
D—SUSPENSION AND FRAME ASSEMBLY
E—SUPPORT ROLLER
F—TAIL PIPE
G—TRACK DRIVING SPROCKET
H—REAR AXLE SHAFT FLANGE
J—CRAB
K—SUSPENSION ARM
L—SUSPENSION FRAME
M—SUSPENSION ROLLER

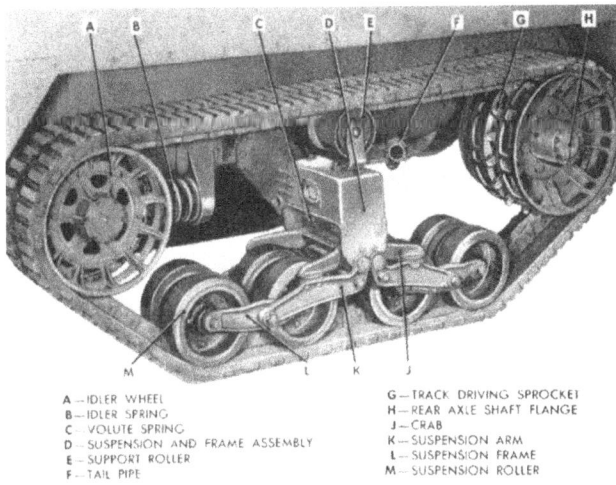

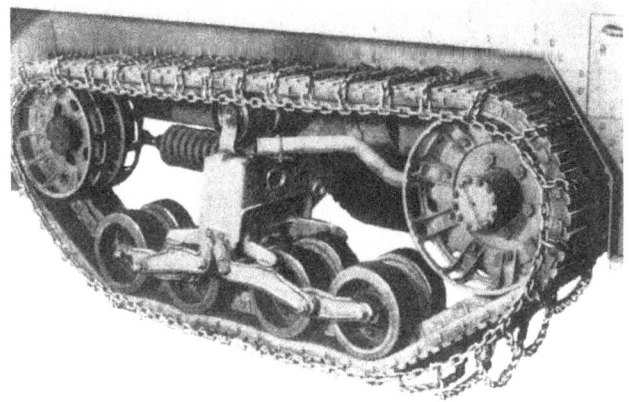

The late production half-track suspension and tracks can be seen above. Note the double coil, spring loaded, idler. At the right, the tracks have been fitted with chains.

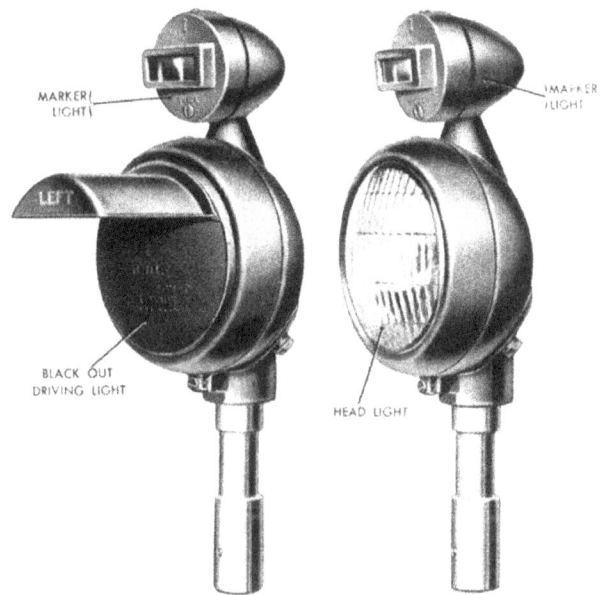

The early headlight with its small blackout marker light is shown above. At the right are two views of the later demountable headlight and blackout marker light. The later type taillights are shown below.

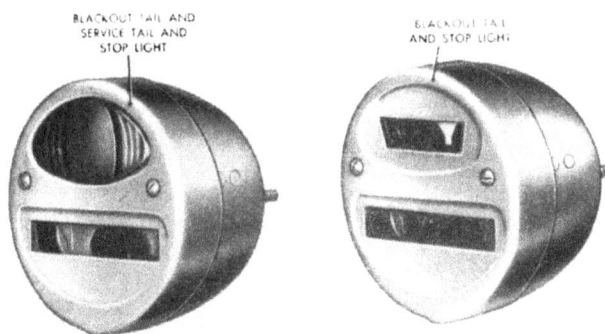

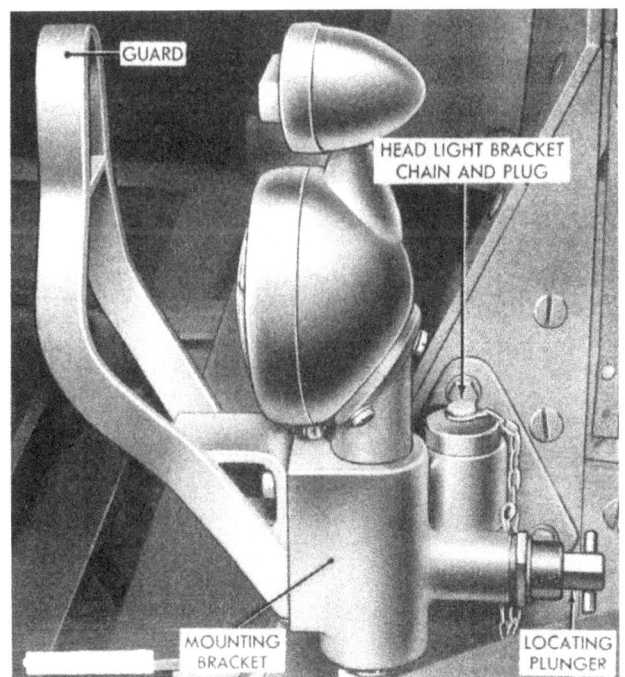

41

Interior arrangements on the half-track car M2 can be seen above with the various stowage compartments open and closed. The vehicle at the top left is fitted with front mounted roller and the one at the top right has the winch. Armament installed on the latter vehicle consists of one .50 caliber M2HB machine gun and one .30 caliber M1919A4 machine gun. The late type demountable headlights are mounted on both vehicles.

The M2 half-track cars below are both fitted with the late type demountable headlights and the double coil, spring loaded, idler. Mine racks have been installed on the M2 at the left and the side stowage compartment is open.

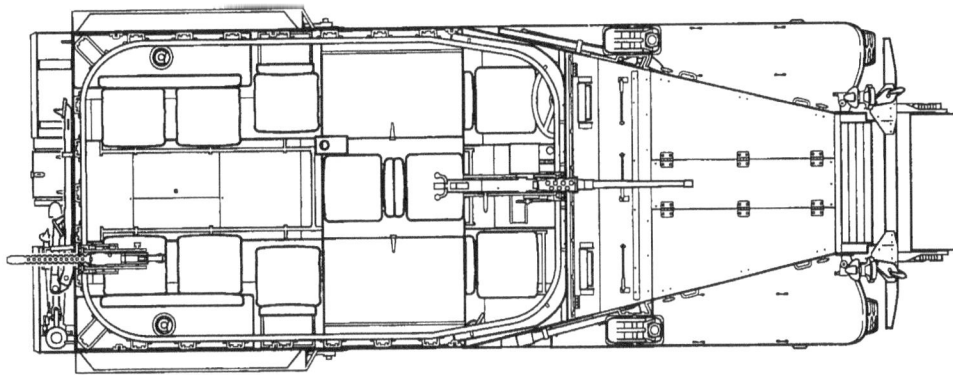

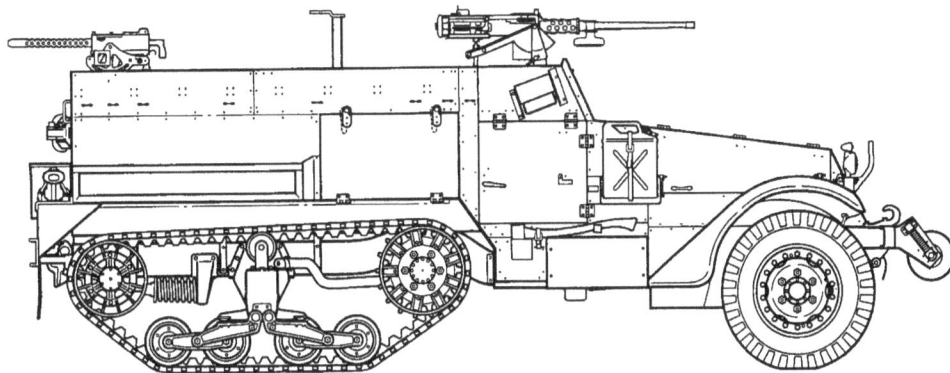

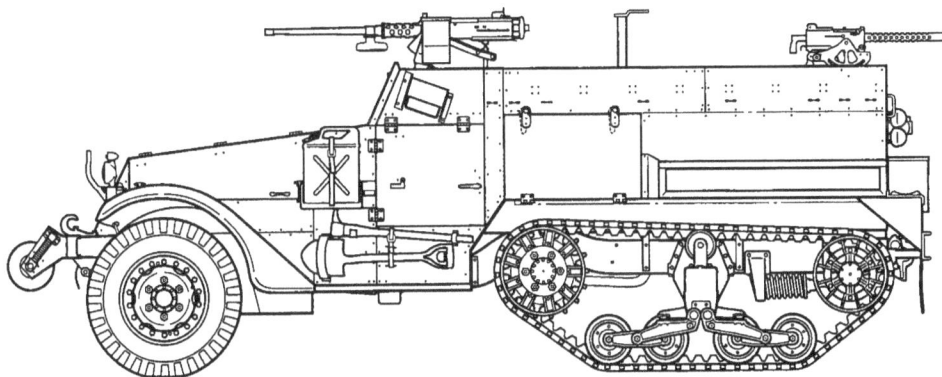

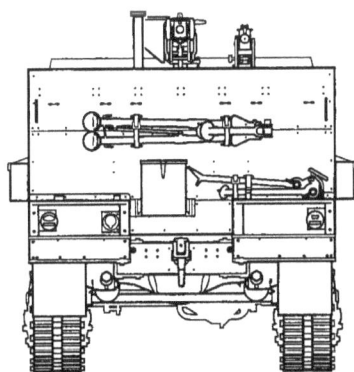

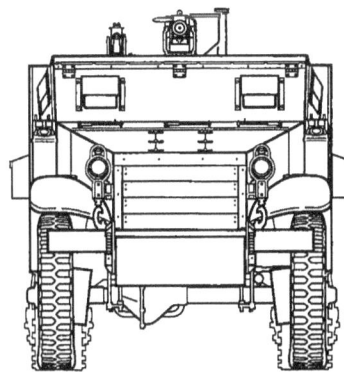

©M. Duplessis

Half-Track Car M2, mid production

The interior of the half-track personnel carrier M3 is visible above with the compartments open and closed. The M3 on the left is equipped with the front roller and the winch is installed on the one at the right. The latter is armed with a single pedestal mounted .30 caliber M1919A4 machine gun. The M3 below is fitted with mine racks and the double coil, spring loaded, idler. All have the late demountable headlights.

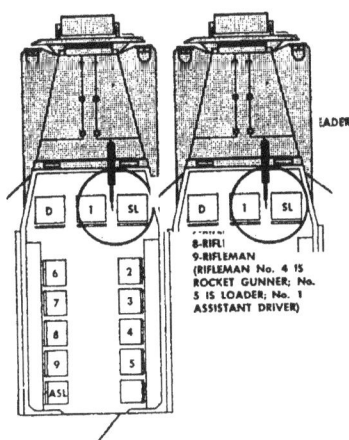

Rifle Squad

8-RIFLEI
9-RIFLEMAN
(RIFLEMAN No. 4 IS
ROCKET GUNNER; No.
5 IS LOADER; No. 1
ASSISTANT DRIVER)

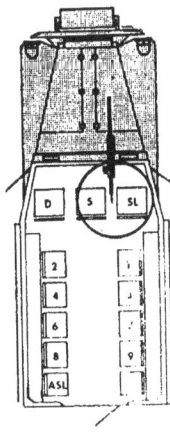

Light Machine Gun Squad

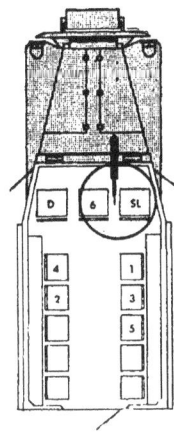

SL-SQUAD LEADER
ASL-ASSISTANT SQUAD LEADER
1-GUNNER
2-GUNNER
3-AMMUNITION HANDLER
4-AMMUNITION HANDLER
5-RIFLEMAN
6-RIFLEMAN
7-RIFLEMAN
8-RIFLEMAN
9-RIFLEMAN
D-DRIVER
(RIFLEMAN No. 7 IS
ROCKET GUNNER; No.
9 IS LOADER; No. 5
IS ASSISTANT DRIVER)

60mm Mortar Squad

SL-SQUAD LEADER
1-GUNNER
2-GUNNER
3-AMMUNITION HANDLER
4-AMMUNITION HANDLER
5-RIFLEMAN
6-RIFLEMAN
D-DRIVER
(RIFLEMAN No. 5 IS
ROCKET GUNNER; No.
6 IS LOADER AND
ASSISTANT DRIVER)

The seating arrangement and armament for the various armored infantry squads are shown in these sketches. Dated 31 January 1945, they are based upon the half-track personnel carrier M3A1. However, the arrangement would be the same with the half-track personnel carrier M3. The vehicle weapons for each type of squad are listed below.

Rifle Squad: (1) .30 caliber M1917A1 MG
(1) 2.36 inch rocket launcher
LMG Squad: (1) .50 caliber M2HB MG
(1) 2.36 inch rocket launcher
60mm Mortar Squad: (1) .30 cal. M1917A1
(1) 2.36 inch rocket launcher
HMG Section: None, but their (2) M1917A1
MGs can be mounted
57mm Gun Squad: (1) .50 cal. M2HB MG or
(1) .30 cal. M1917A1 MG

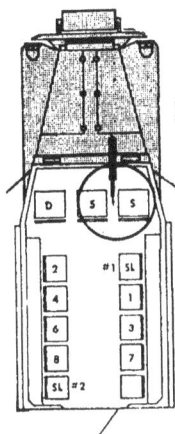

Heavy Machine Gun Section

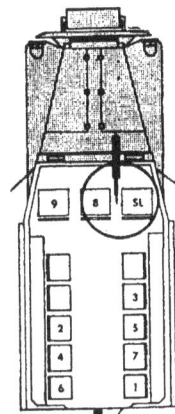

S-SECTION LEADER
SL#1-SQUAD LEADER No. 1
SL#2-SQUAD LEADER No. 2
1-GUNNER
2-GUNNER
3-GUNNER
4-GUNNER
5-AMMUNITION HANDLER
6-AMMUNITION HANDLER
7-AMMUNITION HANDLER
8-AMMUNITION HANDLER
D-DRIVER
(No. 7 IS ROCKET GUNNER;
No. 8 IS LOADER; No. 5
IS ASSISTANT DRIVER)

57mm Gun Squad

SL-SQUAD LEADER
1-GUNNER
2-CANNONEER
3-CANNONEER
4-CANNONEER
5-CANNONEER
6-AMMUNITION BEARER
7-AMMUNITION BEARER
8-AMMUNITION BEARER
9-DRIVER
(AMMUNITION BEARER No.
6 IS ROCKET GUNNER;
No. 7 IS LOADER; No. 8
IS ASSISTANT DRIVER)

Progress in the development of welding techniques for thin armor plate had reached a point by 1941 that it was considered for the fabrication of half-track armor. OCM 16410, dated 16 January 1941, recommended that two welded bodies be procured for the half-track car M2 and one for the half-track personnel carrier M3. When the vehicles were assembled, they were designated as the half-track car M2E1 and the half-track personnel carrier M3E1. The tests at Aberdeen Proving Ground revealed that the welded bodies were less vulnerable to lead splash than the standard bodies with their many joints. Also, there were no cap screws that might be driven into the vehicle by a direct hit. However, the tests did show that the welded homogeneous steel armor was more easily penetrated than the face hardened plate. It was necessary to increase the thickness to achieve adequate protection. In April 1942, the Ordnance Committee authorized welding for the fabrication of thin armor plate and it was to be used by International Harvester Company when they entered half-track production.

On 26 June 1941, OCM 16913 authorized the procurement of a Hercules DWXC diesel engine for installation and test in the half-track car M2. The modified vehicle was to be designated as the half-track car M2E4. The DWXC was a six cylinder, four cycle, diesel engine. Unfortunately, the delivery of the engine was delayed because of a low priority and a new policy directive eliminated the use of diesel engines in U. S. combat vehicles. The M2E4 project was canceled in April 1942 without a vehicle being converted.

Above, the pilot half-track car M2E5 appears at the left and the pilot half-track personnel carrier M3E2 is at the right. Both vehicles, manufactured by International Harvester Company (IHC), were being tested at the General Motors Proving Ground during July 1942.

The great expansion of the War Munitions program after the attack on Pearl Harbor resulted in a large increase in the requirement for half-track vehicles. This increase was far beyond the capacity of the three manufacturers in the program and many of the critical components such as engines and transmissions were in short supply. Fortunately, the truck manufacturing facility of the International Harvester Company (IHC) was available and they could also furnish many of the critical components.

In April 1942, the Ordnance Committee directed that pilot vehicles be procured from IHC for evaluation. These pilots were designated as the half-track car M2E5 and the half-track personnel carrier M3E2 corresponding to the standard M2 and M3 vehicles. Although they used many new components, the performance of the M2E5 and the M3E2 during tests at the General Motors Proving Ground was comparable to the standard half-tracks. The bodies of pilots followed the configuration of the standard half-tracks with the M2E5 having the shorter body of the M2 as well as the side doors for the ammunition storage compartments. However, they could easily be identified by the front fenders. These fenders had an angular profile with a flat cross section. The

early large headlights were mounted on the fenders behind new brush guards.

The production version of the M3E2 was standardized as the half-track personnel carrier M5 by OCM 18370 on 18 June 1942. Standardization of the production model of the M2E5 was approved by OCM 18509 on 10 July 1942 as the half-track car M9.

The production models of the IHC half-tracks differed widely in appearance from M2E5 and M3E2 pilots. Externally, the bodies of both vehicles were identical and were fabricated from homogeneous steel armor. This armor was easily welded and could be formed. The latter resulted in an obvious identification feature of the new vehicles. Smooth round rear corners on both replaced the sharp corners on the M2 and M3 and the side doors of the M2 were eliminated on the M9. Both the M5 and M9 designs had a rear door in the crew compartment. The angular profile of the fenders on the pilots was replaced by a smooth curve, but they retained the flat cross section. The armor plate thickness was increased from $1/4$ inch to $5/16$ inches to partially compensate for the softer homogeneous steel armor. The windshield cover plate was increased to a thickness of $5/8$ inches.

Below are two views of a model of the half-track personnel carrier M5 proposed for manufacture at International Harvester Company (IHC). The model illustrates the smooth welded armor and is fitted with other late features such as mine racks and demountable headlights.

These photographs show the production half-track personnel carrier M5 fitted with the canvas top. The main identification features such as the rounded rear corners and the distinctive front fenders are clearly visible.

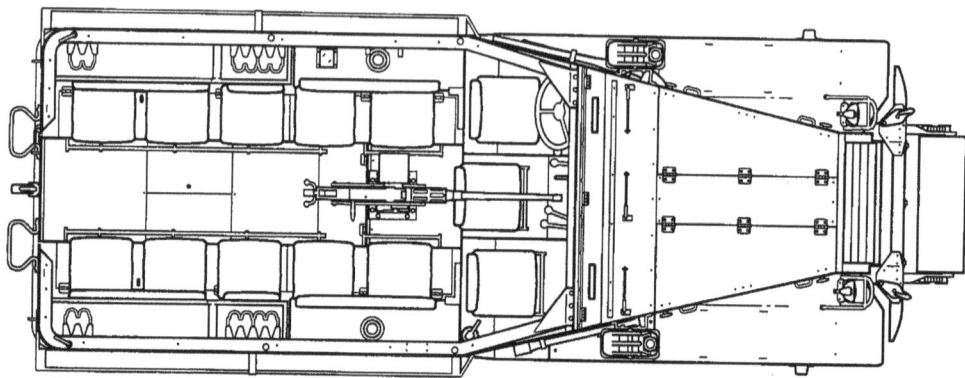

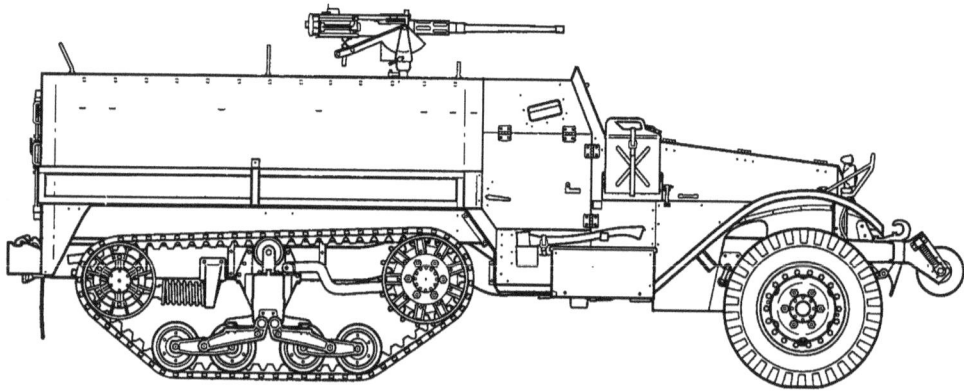

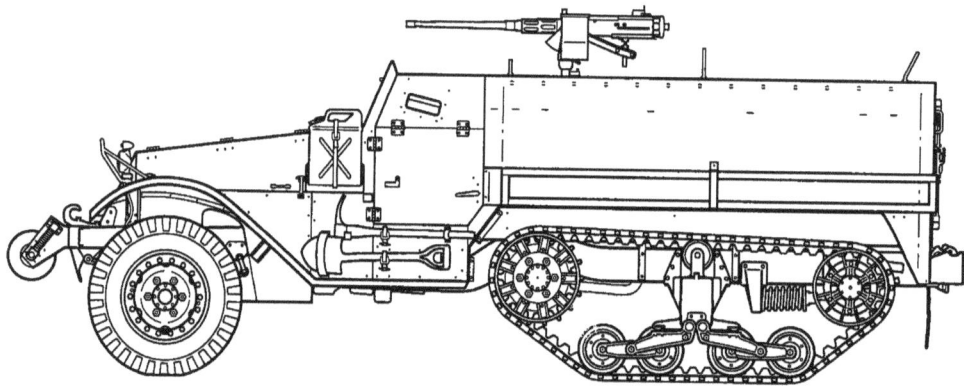

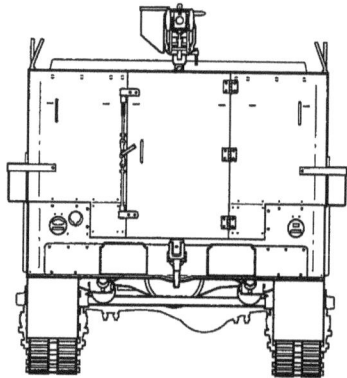

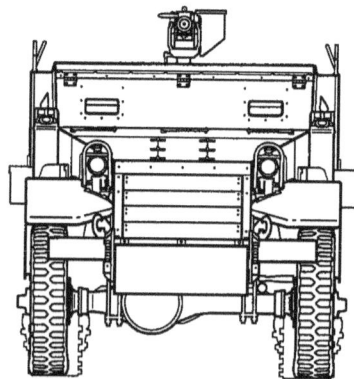

Scale 1:48

Half-Track Personnel Carrier M5

The top views show the interior details of the half-track personnel carrier M5 with the stowage compartments open and closed. Below, the details of the windshield and the armor windshield cover can be seen.

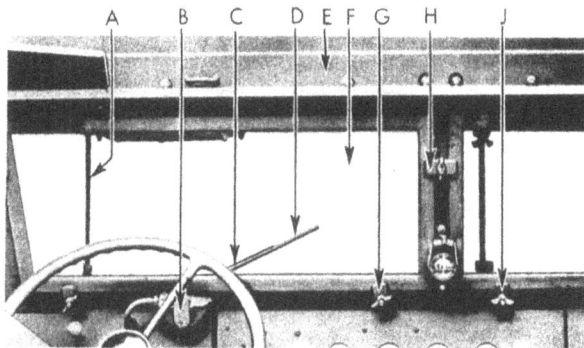

A—WINDSHIELD ARMOR LEFT
      SUPPORT ROD ASSEMBLY
B—WINDSHIELD WIPER ASSEMBLY
C—WINDSHIELD WIPER ARM
D—WINDSHIELD WIPER BLADE
E—WINDSHIELD HEADER

F—LEFT WINDSHIELD, WITH
      GLASS, ASSEMBLY
G—WINDSHIELD HALF CLAMP
H—WINDSHIELD FULL CLAMP
J—WINDSHIELD WING NUT

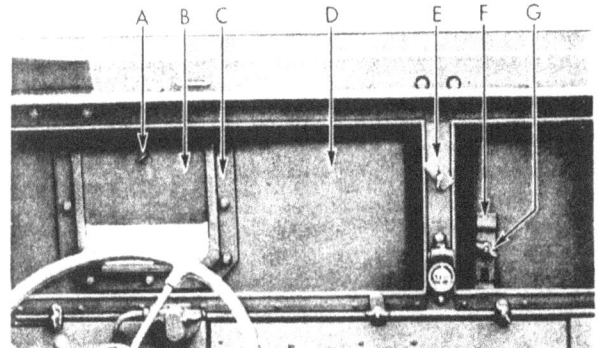

A—WINDSHIELD ARMOR PORT
      HOLE COVER THUMB SCREW
B—WINDSHIELD ARMOR PORT
      HOLE COVER
C—PORT HOLE COVER RETAINER

D—WINDSHIELD ARMOR PLATE
      ASSEMBLY
E—CENTER RETAINER
F—ARMOR PLATE RETAINING CLAMP
G—WING NUT

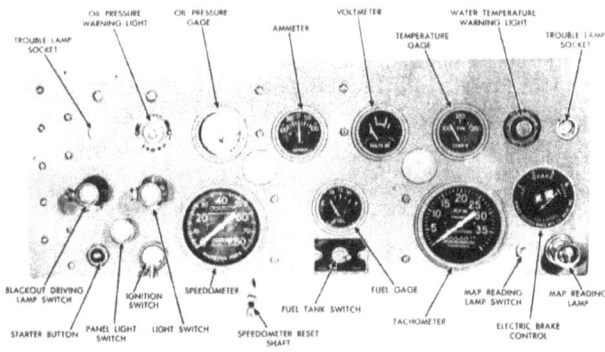

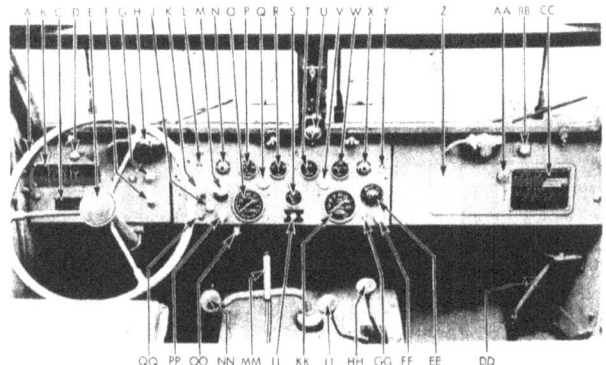

The instrument panel and the driver's controls for the IHC half-tracks are illustrated below. The items in the latter are identified as follows:
A. steering wheel, B. gear shift plate, C. winch operation caution plate, D. vent control, E. horn, F. throttle, G. choke, H. windshield wiper, J. panel light, K. blackout light, L. inspection light socket, M. blackout and service light switch, N. oil pressure warning light, O. speedometer, P. oil pressure gage, Q. panel light cover, R. ammeter, S. fuel gage, T. voltmeter, U. compass, V. panel light cover, W. coolant temperature gage, X. coolant warning light, Y. inspection light socket, Z. map compartment, AA. lock, BB. air cleaner control, CC. registration plate, DD. radiator shutter control, EE. control for trailer brakes, FF. map light, GG. map light switch, HH. front drive shift lever, JJ. transfer case shift lever, KK. tachometer, LL fuel tank selector, MM. hand brake, NN. transmission shift lever, OO. speedometer reset shaft, PP. ignition switch, QQ. Starter.

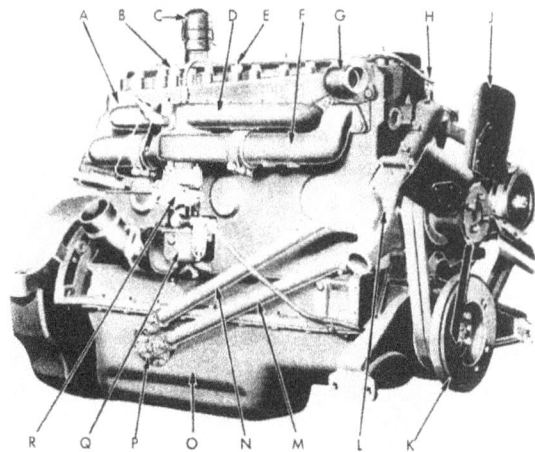

Above are two views of the International Harvester Company Red-450-B engine. The items in the two photographs are identified as follows: Left photo: A. crankcase ventilator, B. oil filler, C. rocker arm cover, D. distributor, E. cylinder breather, F. ignition coil, G. fuel pump, H. starter, J. oil pan, K. oil filters, L. oil level gage, M. generator, N. water pump, O. fan. Right photo: A. cylinder head, B. engine temperature warning unit, C. cylinder head breather, D. intake manifold, E. engine temperature sending unit, F. exhaust manifold, G. thermostat housing, H. crankcase breather, J. fan, K. vibration damper, L. water pump, M. oil cooler outlet, N. oil cooler inlet, O. oil pan, P. oil cooler manifold, Q. carburetor, R. governor

The new half-tracks were powered by the IHC Model RED-450-B six cylinder engine developing 143 horsepower at 2700 rpm. It drove the vehicles through a Spicer designed Model 1856 transmission produced by IHC. The front wheels were fitted with 9:00 x 20 combat tires and heavier front springs were installed. The IHC vehicles benefited from the proving ground tests and field experience of the M2 and M3 half-tracks. This also resulted in the installation of heavier front and rear axles, radiator cross braces, and other improvements indicated by the experience with the earlier vehicles. A new tank type removable instrument panel was provided for the driver. Although the thicker armor and heavier components resulted in a weight increase, the performance of the IHC vehicles remained about the same as the earlier half-tracks.

IHC delivered the first M5 half-track personnel carrier in December 1942. Production continued until September 1943 for a total run of 4625 vehicles. No M9 half-tracks were manufactured as the M9A1 modifications were introduced before production began.

Details of the front roller and the rear bumpers on the half-track personnel carrier M5 can be seen in the two photographs at the right.

Above, the first pilot half-track car M2E6 is at Aberdeen Proving Ground on 3 August 1942 with the original design ring mount for the .50 caliber machine gun.

On 19 May 1942, the Ordnance Committee recommended that the machine gun mounts in the half-track car M2 be changed eliminating the gun rail with the skate mounted machine guns. It was to be replaced by an M32 truck type ring mount over the right front of the vehicle. This concept was approved by OCM 18394, dated 25 June 1942, and the modified vehicle was designated as the half-track car M2E6. Two pilot vehicles were authorized. The first pilot was modified at Aberdeen Proving Ground by installing the 42 inch diameter ring, cradle and carriage from the M32 truck mount supported by three brackets over the right front of the half-track car M2. After test by the Armored Force Board at Fort Knox, it was recommended that the M2E6 be standardized and replace the M2 in production.

However, the second pilot was an even better solution. On this vehicle, the three mounting brackets were replaced by armor plate covering the front and side of the mount providing additional protection to the gunner. The Ordnance Committee adopted the latter design and directed that it be applied to all future production of half-track cars and personnel carriers. The new ring mount was designated as the M49. With the new mount, the vehicles became the half-track cars M2A1 and M9A1 and the half-track personnel carriers M3A1 and M5A1. In addition to the M2 .50 caliber machine gun, armament on the new vehicles included a single M1919A4 .30 caliber machine gun for which three pintle socket mounts were provided, one on each side and one in the rear.

Below, the second pilot half-track car M2E6 is under test at Aberdeen on 1 February 1943. Note the improved protection for the gunner around the .50 caliber machine gun ring mount.

The half-track car M2A1, shown on this and the three following pages, was Ordnance serial number 19496 and it was manufactured by the White Motor Company. These photographs and those on the three following pages were taken at the Engineering Standards Vehicle Laboratory in Detroit, Michigan on 15 February 1944.

The M2A1 replaced the M2 on the production lines at Autocar and White Motor Company and the M3A1 took over from the M3 at Autocar and Diamond T. Total production of new M2A1 half-track cars and new M3A1 half-track personnel carriers was 1643 and 2862 respectively. In addition, 1360 M3A1s were converted from 75mm gun motor carriages M3.

International Harvester Company (IHC) shifted production from the M5 to the M5A1 half-track personnel carrier in October 1943 and it continued until March 1944. A total of 2959 M5A1s were built. The

M9A1 entered production at IHC in March 1943 and a total of 3433 were completed before production was terminated in December of that year. Some documents listed the first 2026 vehicles as M9s, however, they were completed as M9A1s with the ring mount for the .50 caliber machine gun.

The M5, M5A1, and M9A1 half-tracks were classified as Substitute Standard and were allocated to international aid under the Lend-Lease program. Some were assigned to U. S. Army units in the United States for use in training.

Below are additional views of the fully stowed half-track car M2A1 with the canvas top and side curtains installed.

Half-track car M2A1 appears here with the canvas top removed. The vehicle is armed with a .50 caliber M2HB machine gun on the ring mount and a .30 caliber M1919A4 machine gun on a pintle mount. The radio antenna mount has been installed.

Above, the external stowage can be seen on half-track car M2A1, serial number 19496. A full load of 12 M1A1 antitank mines are stowed in the two external racks. The internal arrangement of the vehicle is visible in the top view below. The machine gun armament has been removed in the latter photograph.

Above, details of the winch and demountable headlights are visible at the left and the machine gun tripod mount appears at the right. Below is a view of the driver's compartment on the half-track car M2A1. This is essentially the same as the other late model half-tracks manufactured by White, Autocar and Diamond T.

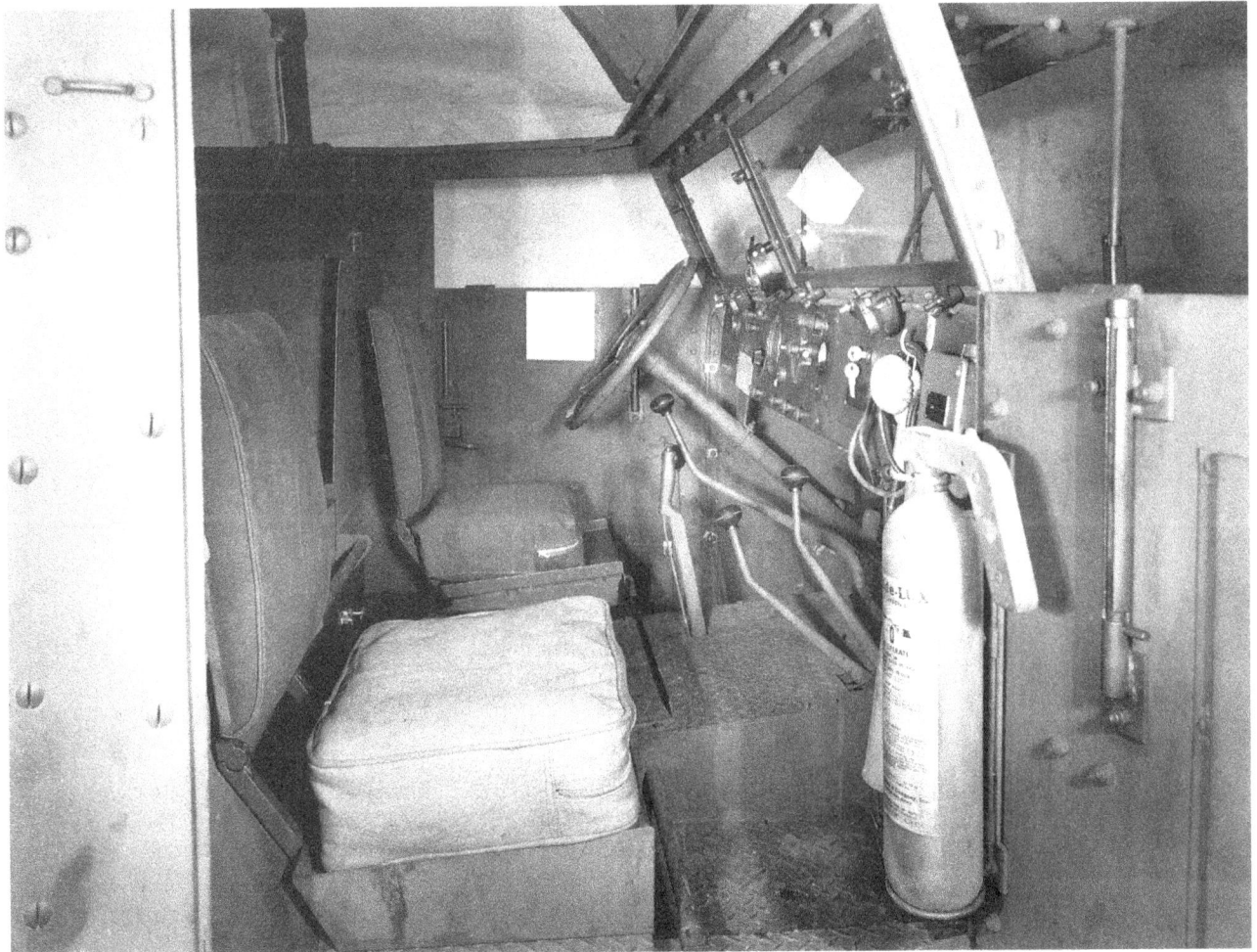

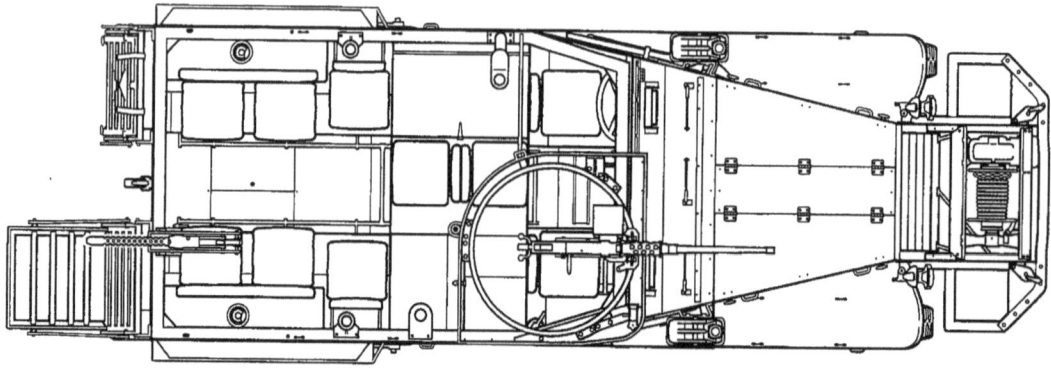

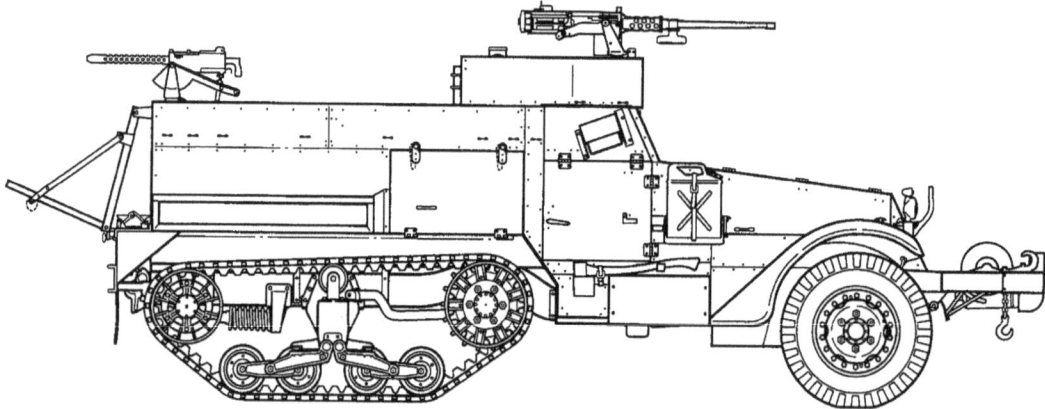

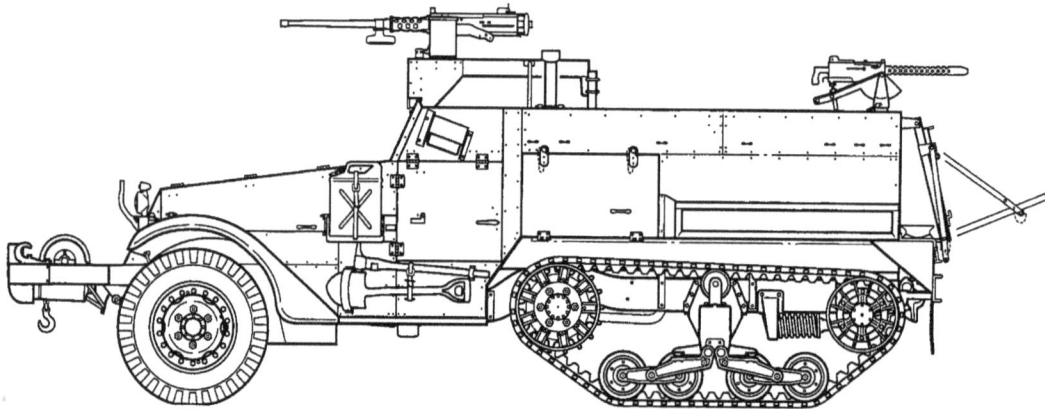

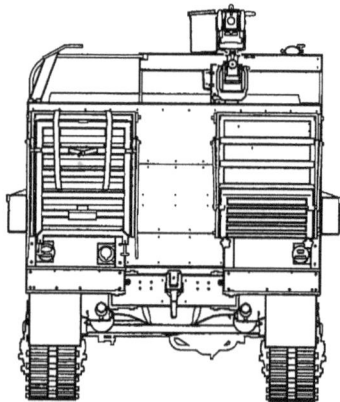

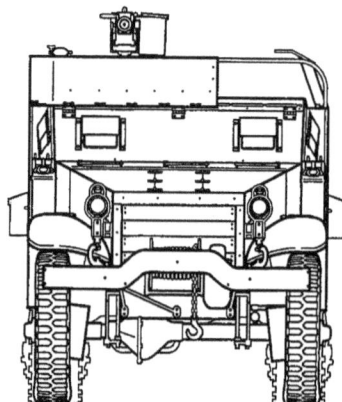

©M. Duplessis

Half-Track Car M2A1

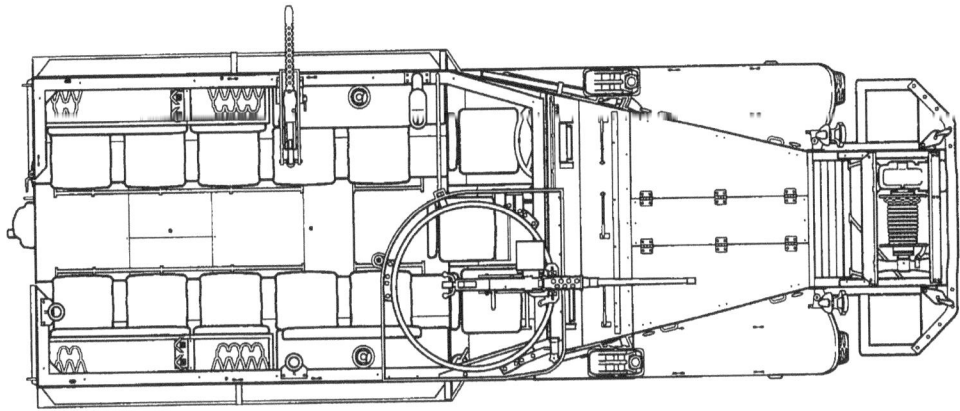

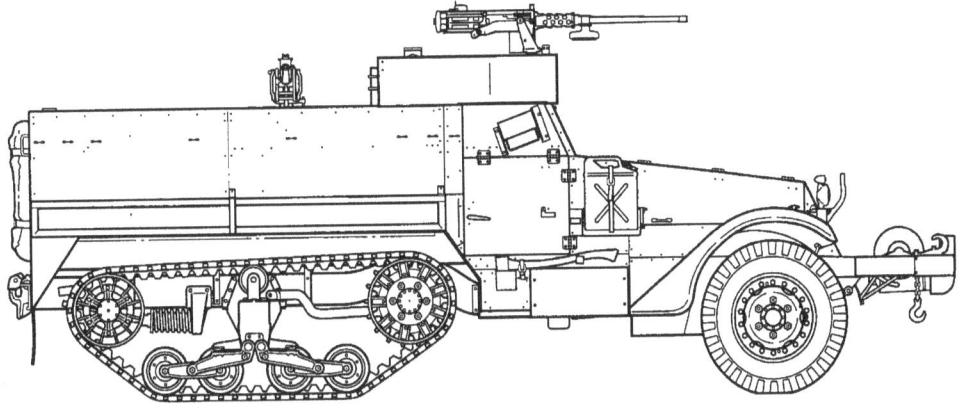

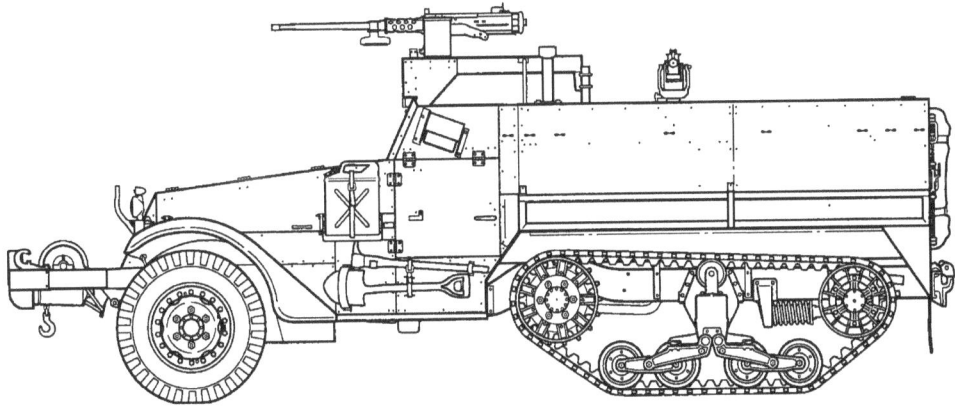

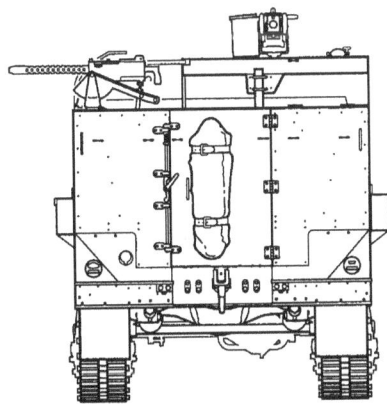

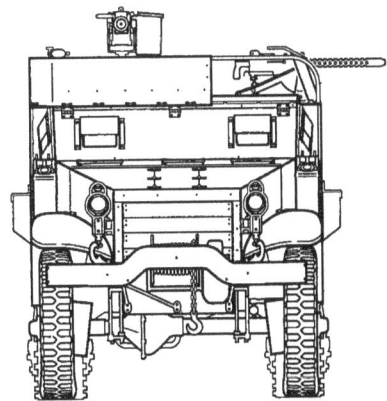

Scale 1:48

©M. Duplessis

Half-Track Personnel Carrier M3A1

The half-track personnel carrier M3A1 appears on this and the following three pages. This vehicle was manufactured by the Diamond T Motor Car Company and had Ordnance serial number 25413. The photographs on this and the next three pages were taken at the Engineering Standards Vehicle Laboratory in Detroit, Michigan on 16 February 1944.

These views of the half-track personnel carrier M3A1 show the fully stowed vehicle with the canvas top removed and the armament of one .50 caliber M2HB machine gun and one .30 caliber M1919A4 machine gun installed. Two radio antenna bases have been mounted.

Above, the 24 M1A1 antitank mines can be seen stowed in the two side racks on the half-track personnel carrier M3A1, serial number 25413. The top view below shows the interior arrangement of the vehicle.

Details on the front and rear of half-track personnel carrier M3A1, serial number 25413, are visible in the two views above. Below is a photograph showing the interior of the vehicle from the rear door looking toward the front.

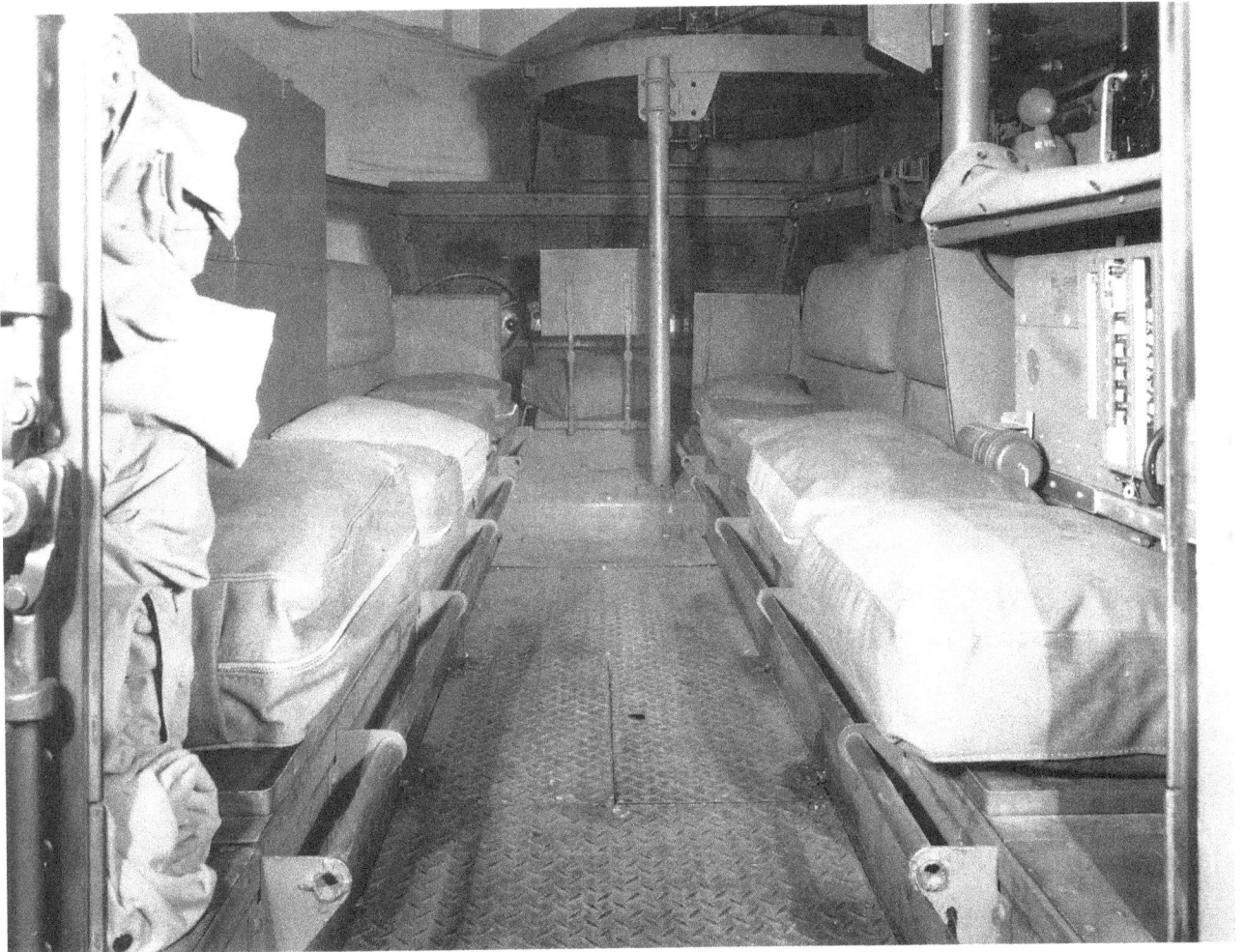

Here are two views of half-track personnel carrier M5A1 manufactured by the International Harvester Company. This vehicle, with Ordnance serial number 18392, was photographed at the Engineering Standards Vehicle Laboratory in Detroit, Michigan on 29 April 1944. In these photographs, it is fully stowed and the canvas top is installed.

The canvas top has been removed and the side curtains are fitted on the half-track personnel carrier M5A1 above at the Studebaker Proving Ground. This photograph was dated 8 November 1943. Note that they have managed to get 13 M1A1 antitank mines in the left mine rack. Below is another view of our old friend, half-track personnel carrier M5A1, serial number 18392, fully stowed with its armament installed.

These photographs and those on the next page show further details of half-track personnel carrier M5A1, serial number 18392, on 29 April 1944. The interior arrangement and stowage are visible in the view below. The round rear corners of the hull and the distinctive fenders of the International Harvester half-tracks are obvious.

Front and rear views of the half-track personnel carrier M5A1 can be seen above. The radiator louvers are fully open. Below at the left, the hood is open revealing the right side of the engine.

The driver's compartment of the half-track personnel carrier M5A1 appears above at the right. Below are views looking forward (left) and toward the rear (right) in the troop compartment of the half-track personnel carrier M5A1. Note the variety of rifles stowed on the vehicle.

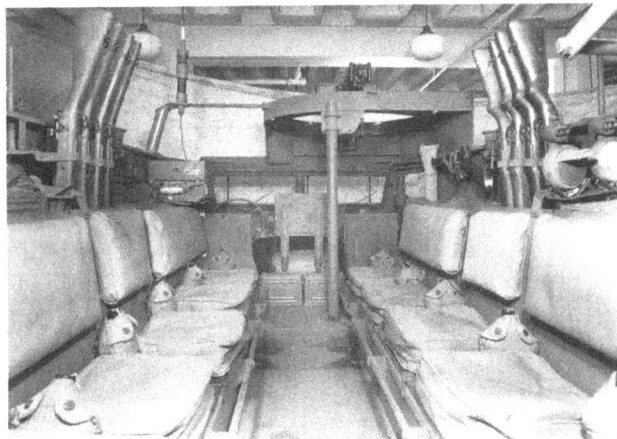

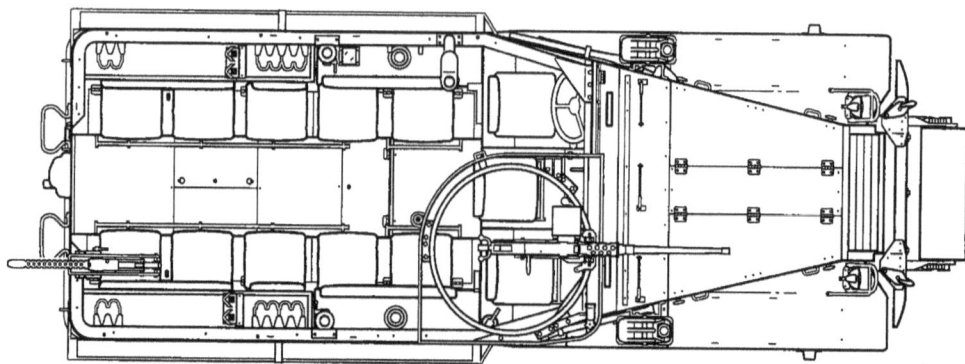

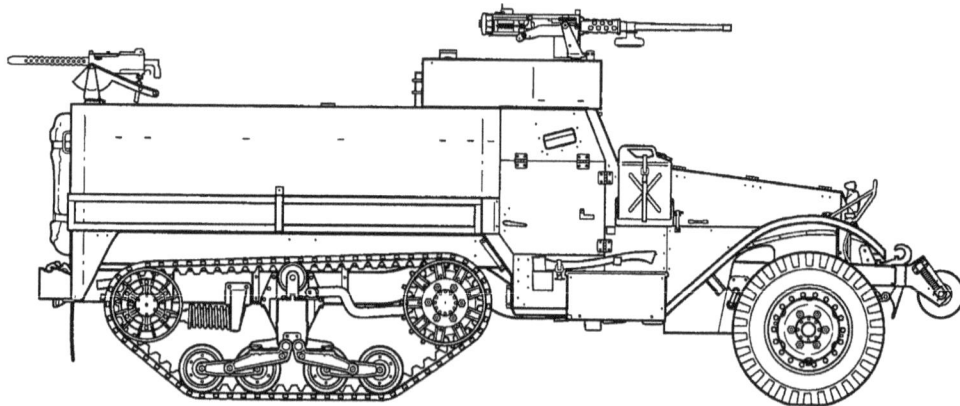

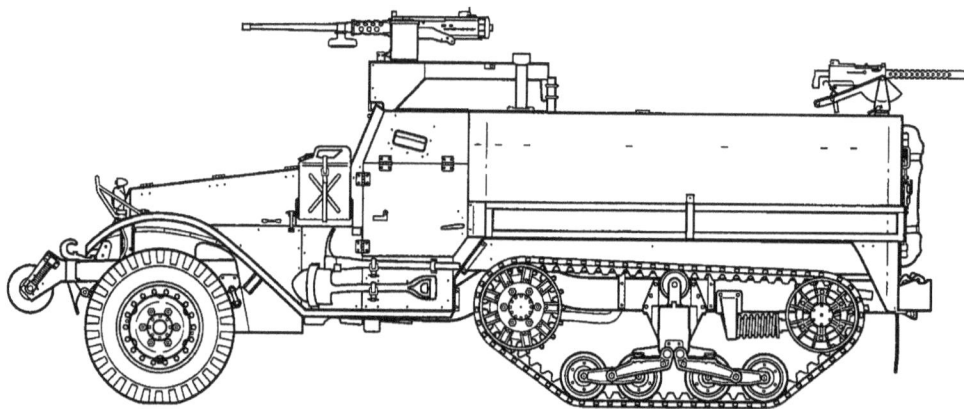

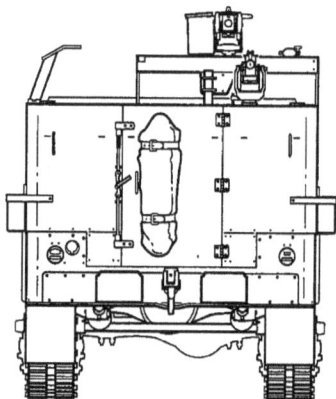

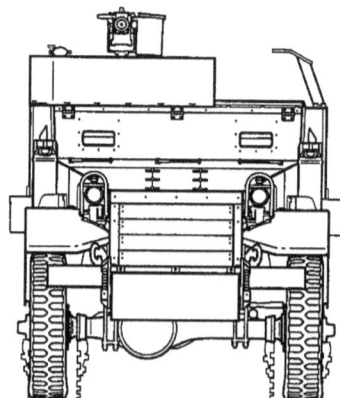

©M. Duplessis

Half-Track Personnel Carrier M5A1

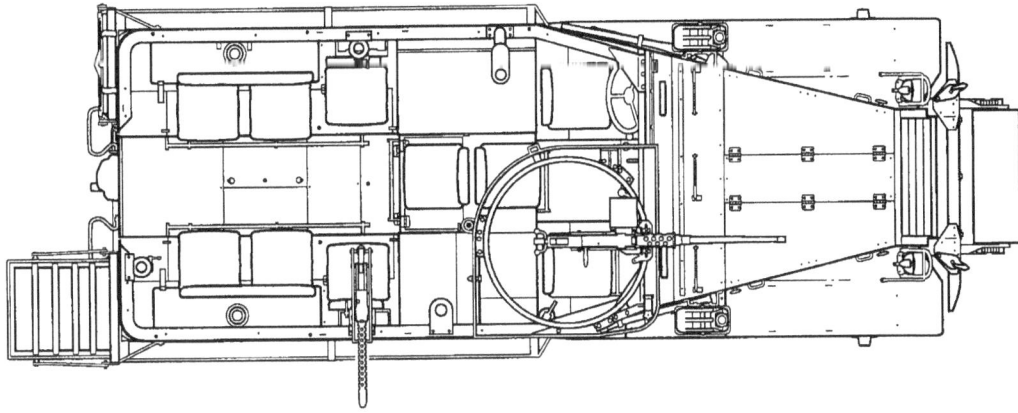

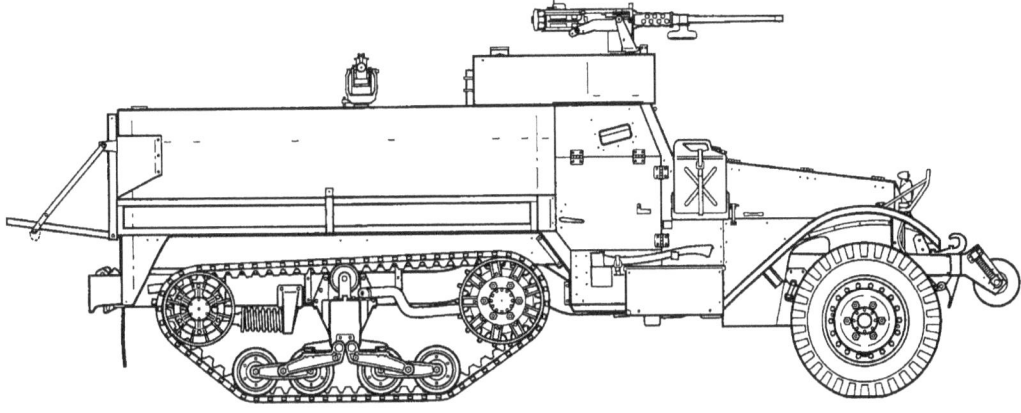

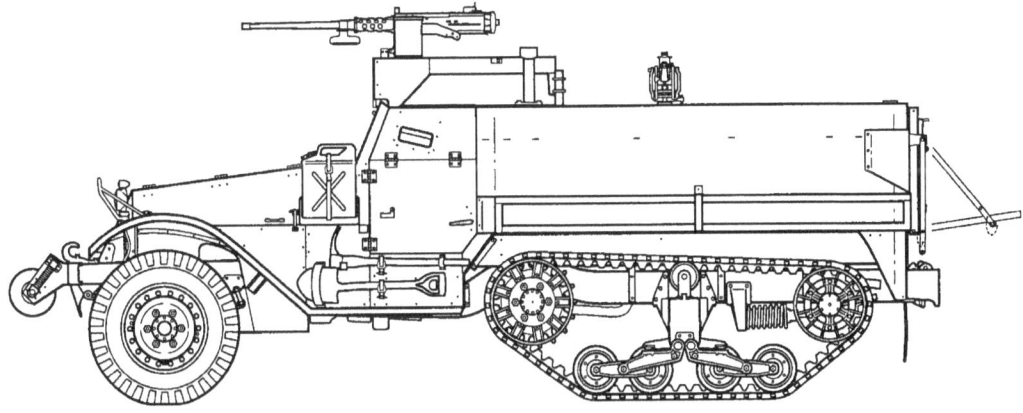

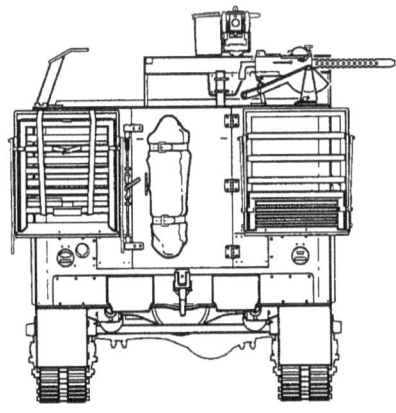

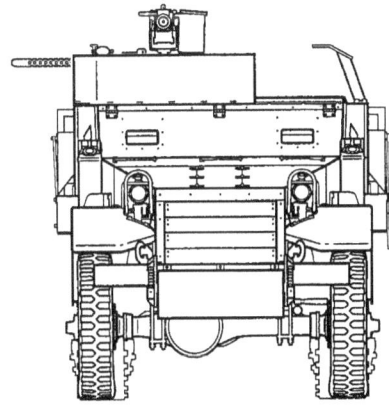

Scale 1:48

Half-Track Car M9A1

The half-track car M9A1 on this and the following three pages was photographed on 11 February 1944 at the Engineering Standards Vehicle Laboratory in Detroit, Michigan. Manufactured by the International Harvester Company, it was assigned Ordnance serial number 2583. The vehicle is fully stowed with the canvas top installed.

Half-track car M9A1 is shown here with the canvas top removed. The armament is installed consisting of one .50 caliber M2HB machine gun on the ring mount and one .30 caliber M1919A4 machine gun on one of the side pintle mounts. The round rear corners of the welded homogeneous steel armor and the characteristic International Harvester fenders are obvious.

The fully stowed half-track car M9A1, serial number 2583, can be seen above and below. A total of 24 M1A1 antitank mines are carried in the side mine racks. The location of the radio and the seating arrangement are visible in the top view.

The front and rear of the IHC half-track car M9A1 appear in the two photographs above. This vehicle is equipped with a front mounted winch in place of the roller. An interior view of the troop compartment appears below looking forward toward the driving compartment. Note the rear door is blocked by the radio installation.

Above are photographs of the half-track car T29 at Aberdeen Proving Ground on 25 June 1943. Note the new rod type side stowage racks and the folding rear stowage racks. The latter are shown in more detail below at the right. This pilot vehicle is equipped with the old style fixed headlights.

A report by the Armored Force Board on Project 302-1, dated 12 December 1942, recommended that the half-track car M2 and the half-track personnel carrier M3 be replaced by a single vehicle based upon the M3. In February 1943, a similar recommendation was made to consolidate the M5 half-track personnel carrier and the M9 half-track car into a single vehicle. Five M3 and three M5 half-track personnel carriers were diverted from production and shipped to International Harvester Company for modification. The proposed replacement for the M2 and M3 was designated as the half-track car T29 and that for the M5 and M9 became the half-track car T31. Both of the pilot vehicles were modified so that they could be adapted for different roles by changes in stowage and items such as radios. The crew was expected to vary from 5 to 12 depending upon the particular application. Both vehicles were armed with the .50 caliber M2 machine gun on the M49 ring mount as on the M3A1 and M5A1. The three pintle sockets, one on each side and one at the rear, also were retained for an M1919A4 .30 caliber machine gun. Other armament depended upon the particular application of the vehicle.

Both of the vehicles were fitted with the new folding stowage racks on the rear and a new side

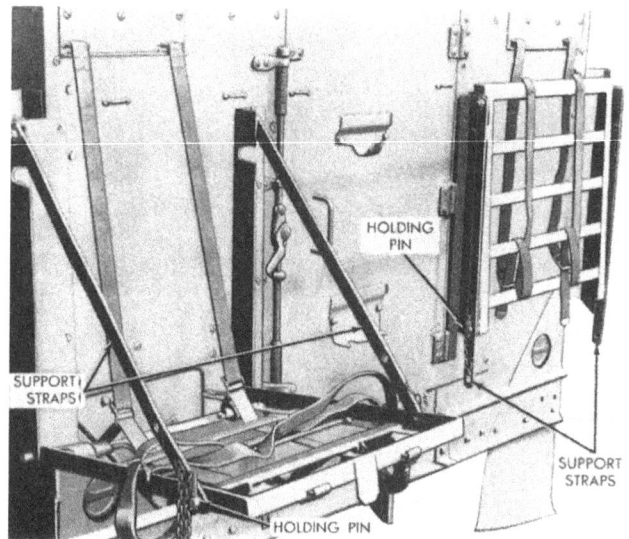

stowage rack. The latter consisted of two rods mounted on the side armor, one above the other, giving the appearance of a ladder. The top rod replaced the loops on the side armor used for securing the canvas top and both could be used to attach stowage to the outside of the vehicle.

Below are two views of the half-track car T31 at Aberdeen on 25 June 1943. The round rear corners of the hull and the fenders characteristic of the IHC vehicles are clearly visible. It also has the new type stowage racks and is fitted with the demountable headlights.

The internal arrangement of the half-track cars T29 (left) and the T31 (right) can be seen above. The armament has not been installed on either of the two vehicles. The SCR 508 radio has been mounted in the T29. Note how the seating varies with different stowage arrangements.

After evaluation, the T29 and T31 were designated as the half-track cars M3A2 and M5A2 respectively. The M3A2 was classified as Standard and the M5A2 was Substitute Standard. Like the earlier International Harvester vehicles, the latter was intended for the Lend-Lease program. Production of the new half-track cars was scheduled for 1 March 1944. However, changing requirements terminated all new half-track production and only the pilot vehicles were completed.

Later in the war, many half-tracks used in training were remanufactured and used to meet the troop requirements after production of new vehicles had ceased.

The vertical armor on the half-track vehicles had always been of concern, particularly when compared with the highly sloped armor on many of the German half-tracks. As a result, a $^3/_{16}$ inch thick spaced armor shell was designed for installation on the M3 half-track. It was highly sloped on the front and sides and a periscope was provided for the driver. However, interest was now shifting to full track carriers for future use and the project was terminated.

Below is an artist's concept of the half-track personnel carrier M3 modified by the installation of the spaced armor shell.

These photographs and those on the next three pages show the pilot half-track car M3A2 at the Engineering Standards Vehicle Laboratory in Detroit, Michigan on 15 September 1944. The vehicle is fully stowed. Note how the canvas top is secured to the upper rod of the new side stowage rack. This vehicle, Ordnance serial number 42977, was manufactured by Autocar.

With the canvas top removed on the half-track car M3A2, the armament of one .50 caliber M2HB machine gun on the ring mount and one .30 caliber M1919A4 machine gun is clearly visible. The latter weapon is installed on the rear pintle mount.

Details of the stowage on half-track car M3A2, serial number 42977, can be seen in the photographs above and below. Note that the bows for the canvas top are stowed behind the new rod type side stowage racks. The seating arrangement is visible in the top view below, but this would, no doubt, vary depending upon the use of the vehicle.

Above are front and rear views of the Autocar half-track car M3A2 pilot. Note that this vehicle is equipped with the demountable headlights and it is fitted with the front mounted roller. The new folding stowage racks can be seen in the rear view. The driver's compartment appears in the view below.

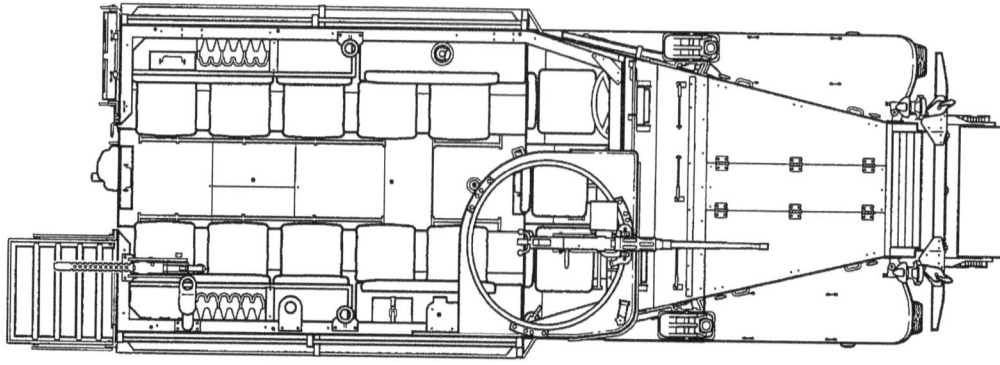

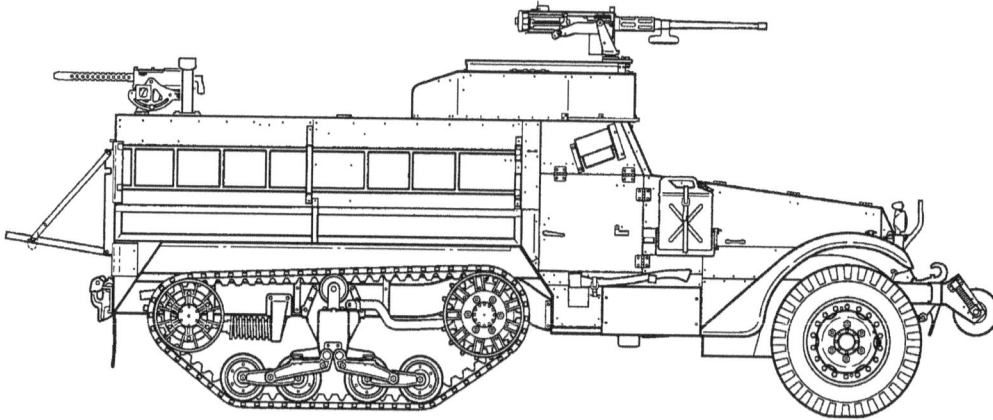

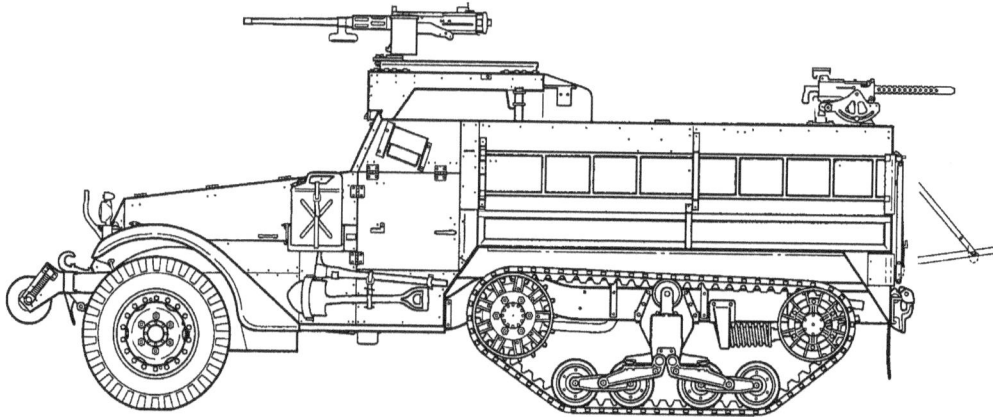

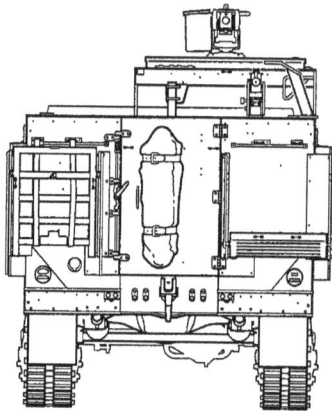

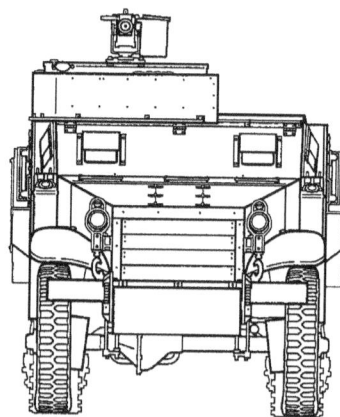

Half-Track Car M3A2

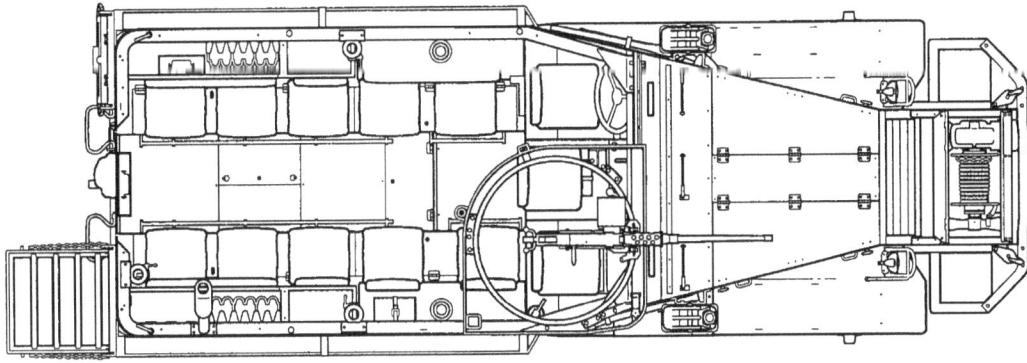

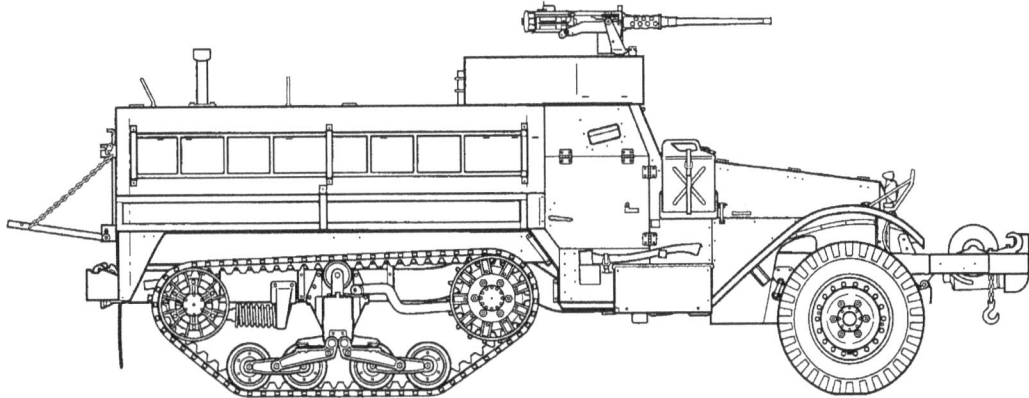

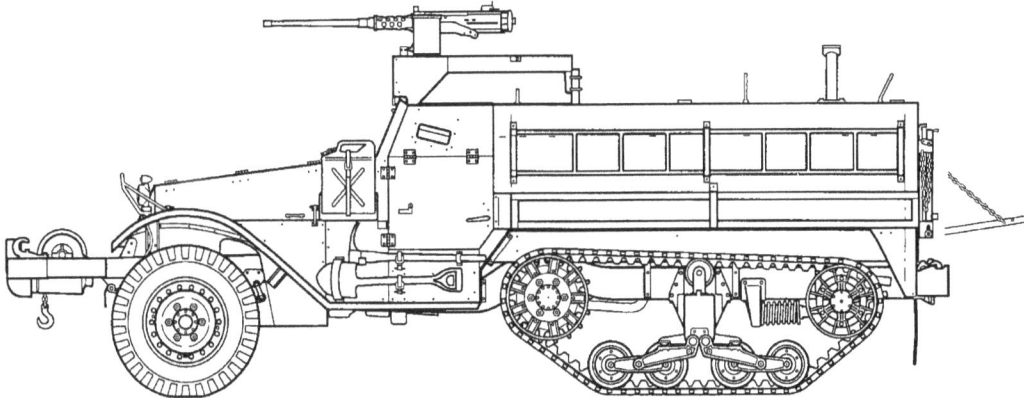

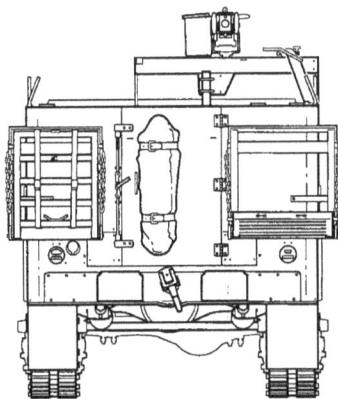

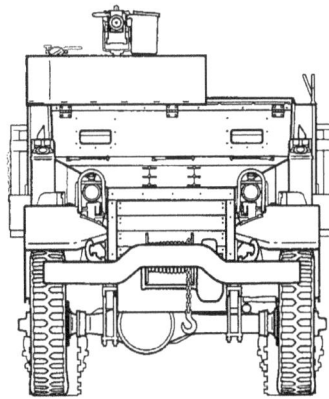

©M. Duplessis

Half-Track Car M5A2

The pilot half-track 81mm mortar carrier M4 above is at Aberdeen Proving Ground on 20 August 1941.

MORTAR CARRIERS

Although the 81mm mortar carrier M4 was standardized in the Fall of 1940 along with the half-track car M2 and the half-track personnel carrier M3, the production pilot of the mortar carrier was not delivered until August 1941. After this single M4 was shipped to Aberdeen Proving Ground, no additional vehicles were received until March 1942, when a total of 278 were completed that month at the White Motor Company. Another 293 M4s were delivered in September and October 1942 bringing the total production to 572. All of the M4s, as well as all of the other half-track mortar carriers during World War II, were manufactured by the White Motor Company.

Additional photographs of the pilot half-track 81mm mortar carrier M4 appear above and below. In the views below, the vehicle is shown with rear door open and closed. The 81mm mortar can be seen through the open door.

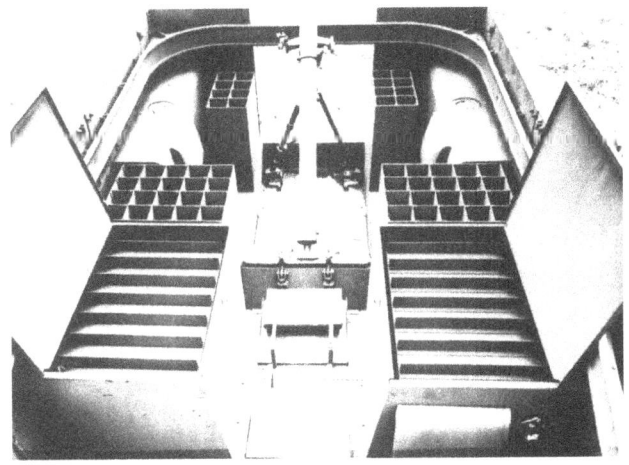

At the top left is a view of the left side ammunition compartment on the half-track 81mm mortar carrier M4 pilot. At the top right, the ammunition racks and the installed 81mm mortar can be seen. These photographs show stowage space for a total of 112 rounds of 81mm ammunition.

As mentioned previously, the M4 was based upon the half-track car M2. A steel support for the mortar base plate was installed with the mortar mounted to fire toward the rear. Initially, the mortar was intended to fire from the vehicle only in emergency situations and the traverse was limited to the 130 mils in the standard mortar bipod mount. Under normal conditions, it was expected that the mortar would be removed from the vehicle and fired from the ground. The hull of the M2 was modified by the installation of a rear door on the M4. However, the gun rail was retained with the skate mounted machine guns so that it was necessary to crawl under the gun rail to use the rear door. The M4 retained the two ammunition compartments with the top and side doors as on the M2, but the compartments were rearranged so that each could stow 28 rounds of 81mm mortar ammunition. On the pilot M4, 56 mortar rounds were stowed in open top racks bringing the total 81mm ammunition stowage to 112 rounds. This resulted from the early concept of using the half-track only for transportation and firing the weapon from the ground. At one point, it was expected to carry 126 rounds of ammunition, although the stowage space for the additional rounds is not obvious. After tests at Aberdeen, two eight round open top racks were removed to provide space for the mortar crew reducing the ammunition stowage to 96 rounds for the production vehicles.

Below, an early production half-track 81mm mortar carrier M4 is at Aberdeen Proving Ground on 25 June 1942. With the canvas top installed, the external appearance of the M4 is the same as that of the half-track car M2 except for the rear door.

In these two photographs, the canvas top has been removed from the early production half-track 81mm mortar carrier M4. In the upper view, the door armor flaps are closed, but the windshield armor cover is open. Below, the door armor flaps are also open.

As can be seen in the photograph above, the early production 81mm mortar carrier M4 retained the stowage space for 112 rounds of 81mm ammunition. The 81mm mortar appears below at the left through the open rear door. Note how the rail for the skate mounted machine guns blocks the rear door.

The SCR-510 radio is installed in the half-track 81mm mortar carrier M4 above at the right. Below are two views of the later production half-track 81mm mortar carrier M4. Note that the 81mm ammunition stowage has been reduced to 96 rounds.

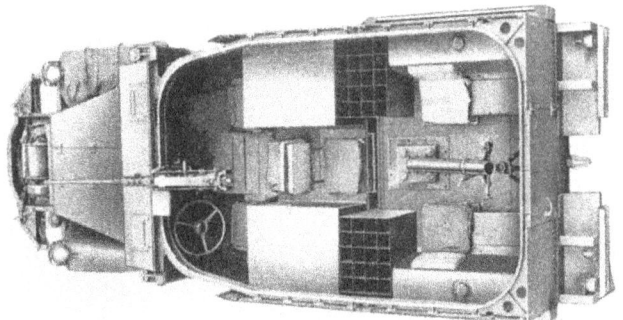

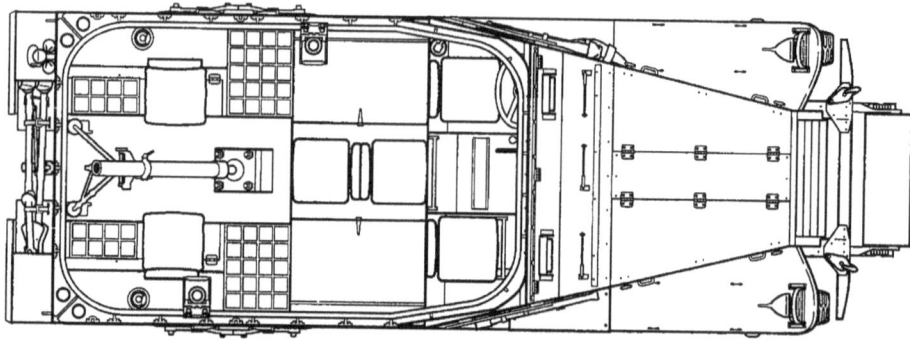

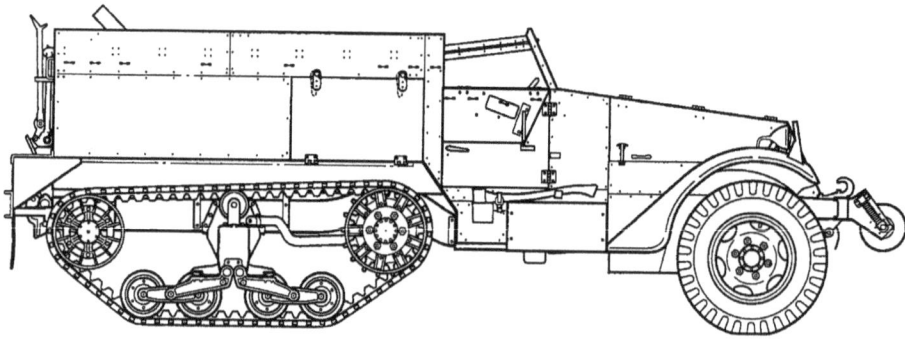

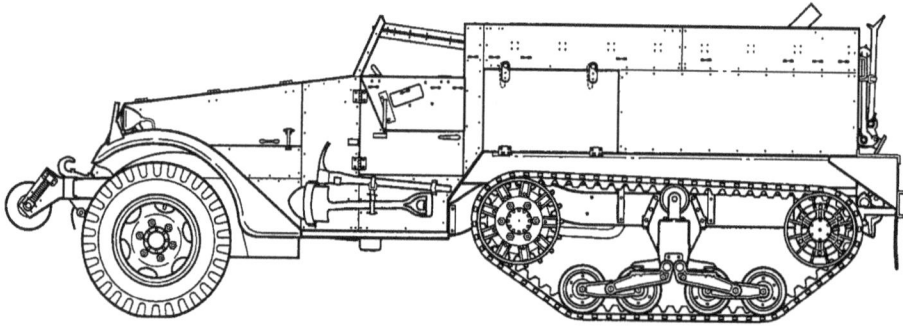

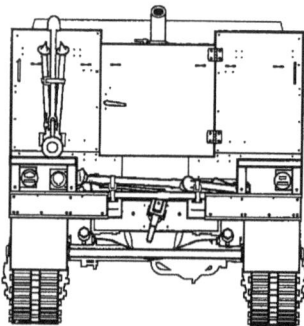

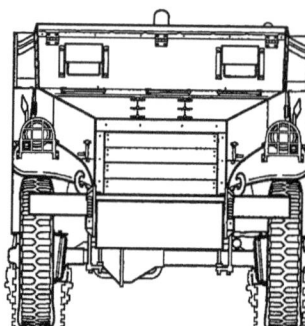

Scale 1:48

Half-Track 81mm Mortar Carrier M4, early production

The half-track 81mm mortar carrier M4A1 appears in the photographs on this and the next two pages at the Engineering Standards Vehicle Laboratory in Detroit, Michigan on 9 February 1944. This vehicle carried Ordnance serial number 1945.

When the new mortar carrier reached the troops, it became obvious that the mortar should be able to fire from the vehicle under all conditions. Since the limited traverse delayed the alignment of the mortar with the target, a traversing fixture was installed to increase the range of traverse to 600 mils in 100 mil increments. Fine adjustments were still made on the bipod mount. The socket plate assembly had to be raised $7^3/_8$ inches to accommodate the new traversing arc. On 28 January 1943, the Ordnance Committee recommended the standardization of the modified vehicle as the 81mm mortar carrier M4A1. A new sight with 6400 mil traverse and an elevation scale in mils was standardized as the M6 for use on the M4A1 mortar carrier. If any of the M4 mortar carriers in the field could not be upgraded because of their location, they were reclassified as Limited Standard.

Production of new M4A1 81mm mortar carriers began at the White Motor Company with the delivery of 100 vehicles in May 1943. The production run continued until October of the same year for a total of 600 M4A1 81mm mortar carriers.

All of these photographs show the M4A1 fully stowed with the canvas top and side curtains installed.

The canvas top and side curtains have been removed from the half-track 81mm mortar carrier M4A1 in these two photographs. The 81mm mortar is visible and the .30 caliber M1919A4 machine gun has been installed on the skate mount

Half-track 81mm mortar carrier M4A1, serial number 1945, appears in all of these views. This is a late production vehicle with the double coil spring on the track idler. By reversing some of the mines, they have managed to stow a total of 14 M1A1 antitank mines in the two side racks. The photograph below shows the interior arrangement of the M4A1. Note the 81mm ammunition stowage remains at 96 rounds. The traversing arc on the new mortar mount is clearly visible.

The interior stowage and crew positions in the half-track 81mm mortar carrier M4A1 can be seen in the views above and at the right. These photographs were taken at Fort Knox. Below is another view of M4A1, serial number 1945, in Detroit on 9 February 1944.

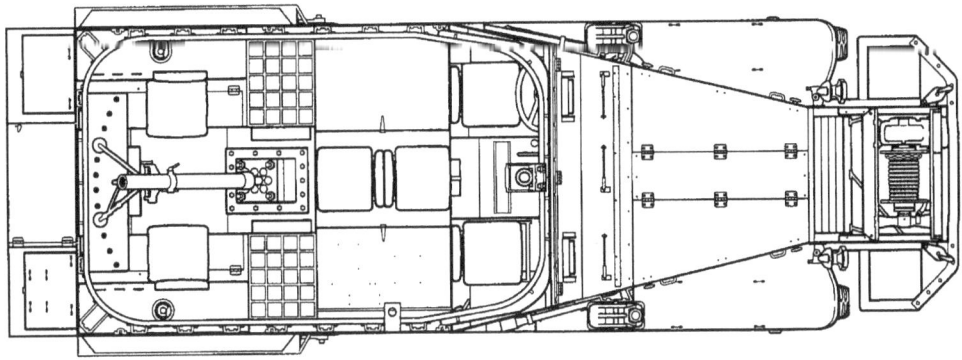

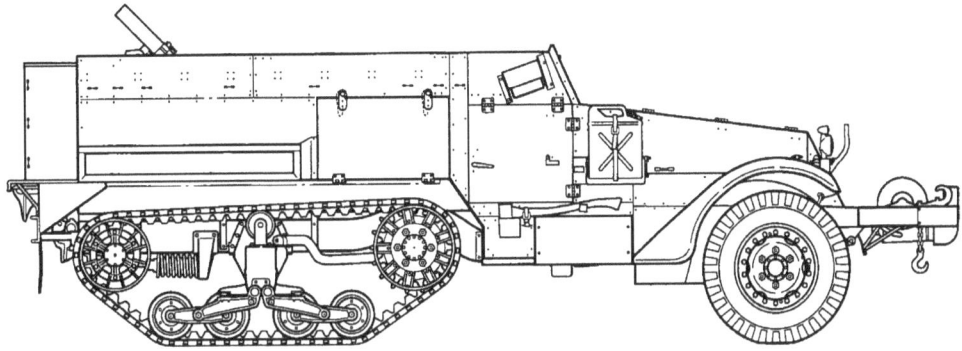

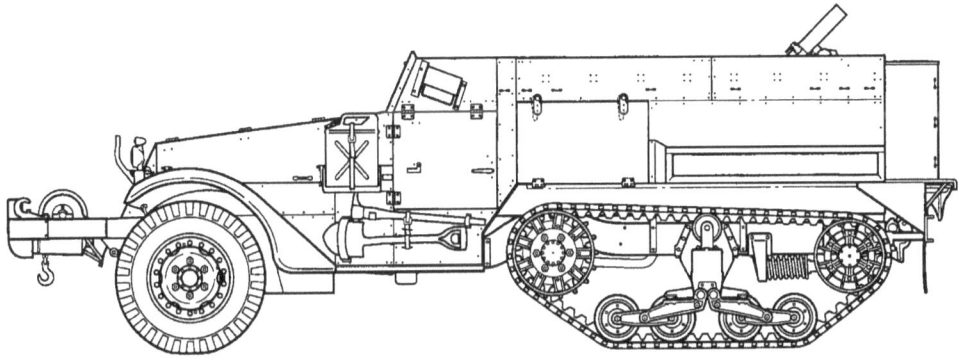

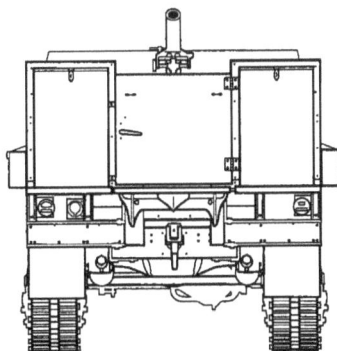

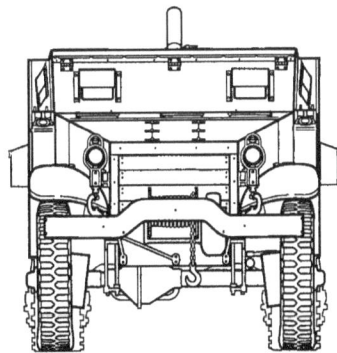

Scale 1:48

Half-Track 81mm Mortar Carrier M4A1

The half-track 81mm mortar carrier M4 above was modified by the 2nd Armored Division so that the mortar fired in the forward direction. Note that the 81mm ammunition stowage has been reduced and the rear door is completely blocked. At the right is the half-track 81mm mortar carrier T19 at Aberdeen Proving Ground on 8 June 1943.

Field experience soon indicated that it would be preferable to have the mortar mounted in the vehicle so that it fired toward the front. New studies of such a mortar carrier based upon the half-track personnel carrier M3 were begun by the Armored Force Board. However, the 2nd Armored Division did not wait for a new development program to provide them with an improved mortar carrier. They converted their M4 carriers by remounting the mortar so that it fired toward the front of the vehicle. These modified M4s operated successfully until the end of the war in Europe.

The new vehicle with the forward firing mortar recommended by the Armored Force Board was designated as the half-track 81mm mortar carrier T19 by Ordnance Committee action in October 1942. Based upon the half-track personnel carrier M3, it was manned by a crew of six men. The pilot T19 was built by the White Motor Company and shipped to Aberdeen Proving Ground for tests which ran from April to July 1943. The Proving Ground recommended some minor modifications including the addition of stiffeners to the pedestal mount for the .50 caliber machine gun to reduce dispersion. With these changes, the T19 was standardized as the half-track 81mm mortar carrier M21 by OCM 20846 dated 7 June 1943. A total of 110 M21 mortar carriers were delivered by the White Motor Company starting in January 1944 and ending in March of the same year. These late production vehicles could be identified by the new rod type side stowage racks.

The half-track 81mm mortar carrier T19 appears at the right. The mortar is aimed forward and a .50 caliber M2HB machine gun is installed on a pedestal mount.

These photographs show the half-track 81mm mortar carrier M21 at the Engineering Standards Vehicle Laboratory on 28 April 1944. This vehicle, Ordnance serial number 25, was manufactured by the White Motor Company.

The half-track 81mm mortar carrier M21 shown here incorporates the latest production features such as the new rod type side and folding rear stowage racks. The 81mm mortar is aimed forward and the secondary armament consists of a single pedestal mounted .50 caliber M2HB machine gun.

The interior stowage of the half-track 81mm mortar carrier M21, serial number 25, can be seen below. The vehicle is equipped with a winch. Note the new 81mm ammunition stowage arrangement.

The interior of the half-track 81mm mortar carrier M21 can be seen above looking toward the rear. Note the extra 81mm rounds stowed alongside the mortar mount. A view looking toward the front appears at the right. The mortar base plate for ground use is stowed on the rear door below compared to the rear hull in the later photograph at the right.

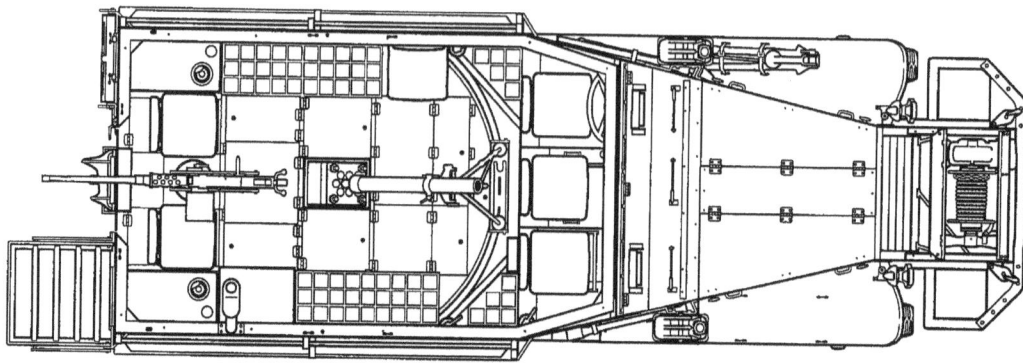

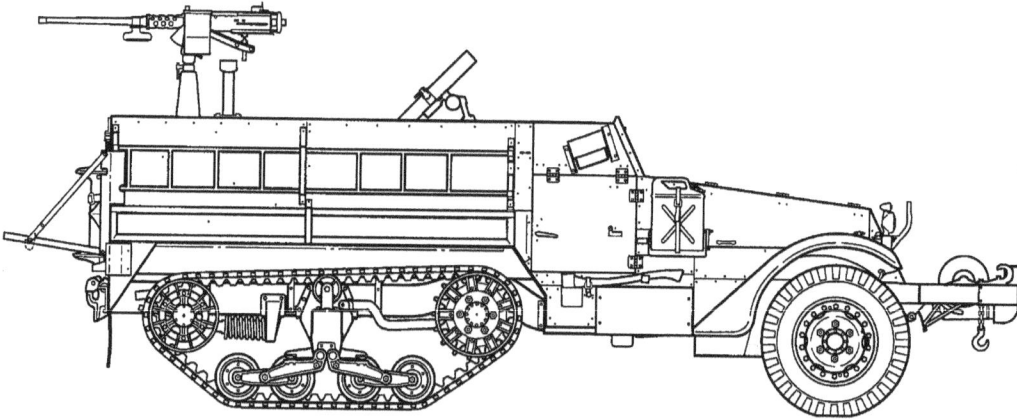

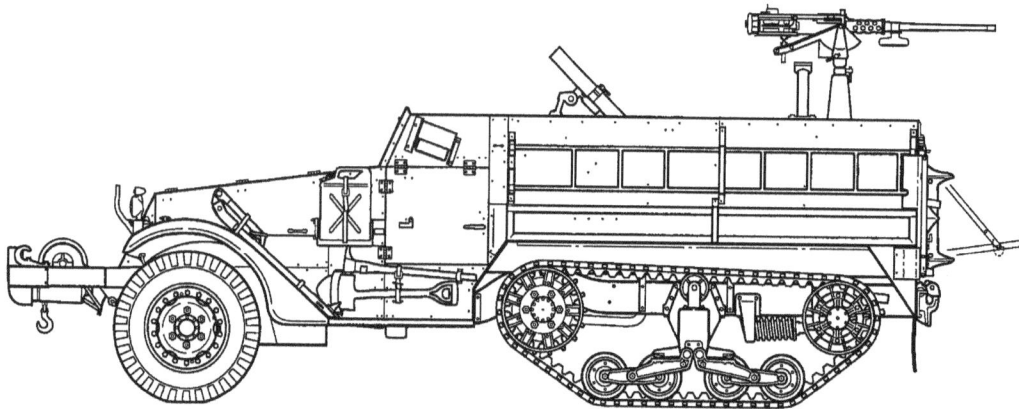

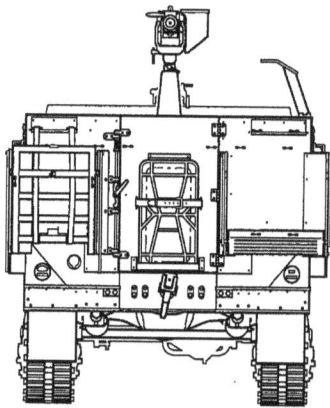

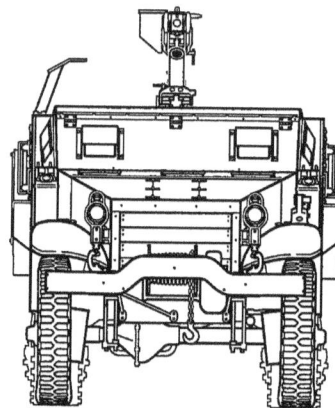

Scale 1:48

Half-Track 81mm Mortar Carrier M21

The half-track 81mm mortar carrier M21 is shown in the photographs on this page during its evaluation at Fort Knox. The opening for the mortar in the canvas top can be seen above and at the left below. Below at the right, the rear view shows the mortar ground base plate still stowed on the rear door. No doubt the door was somewhat heavy to move with the addition of the 45 pound base plate.

Below are two views of the half-track 81mm mortar carrier M21 with the full crew on board. At the bottom right, the loader is about to drop the 81mm round into the mortar tube and another member of the crew is manning the pedestal mounted .50 caliber M2HB machine gun.

In response to a request from the Chief of the Chemical Warfare Service, studies were initiated at Aberdeen Proving Ground in December 1942 to investigate the installation of the 4.2 inch mortar on the half-track. The first tests were made by replacing the 81mm mortar on the M4 carrier with a 4.2 inch M2 chemical mortar. Despite the use of a two inch thick layer of rubber under the base plate, the firing tests caused considerable damage to the supporting structure of the vehicle. The program was then switched to using a half-track personnel carrier M3A1 with a reinforced chassis. This vehicle, designated as the 4.2 inch mortar carrier T21, retained the ring mounted .50 caliber machine gun and installed the mortar firing toward the rear. Although the initial firing tests were successful, subsequent firing caused severe damage to the half-track chassis which would require a complete rebuild. At this point, the specifications for the new vehicle were changed requiring the mortar to be aimed toward the front of the vehicle with a greater range of traverse. OCM 21810, dated 14 October 1943, designated the new vehicle as the 4.2 inch mortar carrier T21E1. The pilot T21E1 was manufactured by the Autocar Company incorporating the latest half-track design features. With the forward firing mortar, the machine gun ring mount was eliminated and the .50 caliber weapon was relocated to a rear pedestal mount.

Although the firing tests were satisfactory, interest had now shifted to the development of a full track 4.2 inch mortar carrier. OCM 27124, dated 29 March 1945, recommended the termination of the T21E1 project.

The half-track 4.2 inch mortar carrier T21 is shown above and below at the right. The mortar is aimed toward the rear and the vehicle retains the armament of the half-track personnel carrier M3A1 with the .50 caliber M2HB machine gun on the ring mount and a pintle mounted .30 caliber M1919A4 machine gun.

The half-track 4.2 inch mortar carrier T21E1 below is based upon the half-track personnel carrier M3 with the mortar aimed forward. The secondary armament now consists of a single pedestal mounted .50 caliber M2HB machine gun.

The original pilot 75mm gun motor carriage T12 is shown here and at the bottom of the page at Aberdeen Proving Ground on 21 July 1941. Note the deflection of the suspension under the load.

TANK DESTROYERS

The effective employment of German armored forces was a subject of serious study by the U. S. Army after the fall of France in 1940. In early 1941, General George C. Marshall commented:

" It occurs to me that possibly the best way to combat a mechanized force would be to create anti-mechanized units on self-propelled mounts, with emphasis on visibility (for the gunner), mobility, heavy armament, and very little armor".

In 1941, there was little available in the way of guns or suitable chassis for such a self-propelled weapon. The 75mm gun M1897A4 was one weapon that existed in sufficient numbers for such an application. This was the modernized version of the French 75 adopted by the U.S. Army during World War I. At the same time, the half-track personnel carrier M3 was just coming into full production. A conference at Aberdeen Proving Ground on 25 June 1941 between representatives of the Ordnance Department and the Assistant Chief of Staff G3 initiated a project to mount the 75mm gun on the M3 as an expedient tank destroyer. This would make good use of the existing stocks of the old 75mm field gun and it might provide a suitable weapon for a new tank destroyer unit during the maneuvers in the Fall of 1941.

A team led by Major (later Colonel) Robert J. Icks rapidly assembled a pilot vehicle at Aberdeen. Designated as the 75mm gun motor carriage T12, it served as a guide for the Autocar Company which received an order for 36 production vehicles. This was shortly increased to 86 and they were all delivered in August and September 1941. The first 36 gun motor carriages were assigned for field testing at Aberdeen and to the 93rd Antitank Battalion, one of the early tank destroyer units. The remaining 50 vehicles were rushed to the Philippines where they served in the defense of the islands against the Japanese. It obviously was a rare occurrence for a new combat vehicle to be in action approximately six months after its conception.

The windshield armor cover is shown open and closed in the two photographs above. These were taken at Aberdeen Proving Ground on 5 August 1941. Note that the gun mount retains the original shield from the 75mm field gun.

Modification of the half-track personnel carrier to the gun motor carriage included removal of the glass windshield and the armored windshield cover was hinged at the bottom instead of at the top so that it folded down on top of the hood. A notch was cut into the top of the armored cover to provide clearance for the barrel of the 75mm gun when it was locked in the travel position. The seats, sub-floor, and gun racks were removed from the M3 and the fuel tanks were relocated on each side at the rear. A steel base was mounted on the half-track frame to support the upper parts of the standard M2A3 gun carriage. The gun was aimed forward over the hood with a total traverse of 40 degrees (19 degrees left and 21 degrees right). The elevation ranged from +29 degrees to almost -10 degrees. The converted mount was standardized as the 75mm gun mount M3. The shield for the M2A3 carriage was retained as well as the standard sight. A pedestal mount was installed for a .50 caliber machine gun. A new sub-floor was added with ten four round boxes for 75mm ammunition and one box for .50 caliber ammunition. An additional 19 rounds of 75mm ammunition were stowed below the gun mount.

At the right, the upper photograph shows the rear of the gun mount on the half-track 75mm gun motor carriage T12. Note the original shield and the stowage space for 19 rounds of 75mm ammunition beneath the mount. The bottom photograph shows the rear of the 75mm gun motor carriage on 21 July 1941. Note the early taillights on this vehicle.

Above at the left, the half-track 75mm gun motor carriage T12 is still fitted with the original shield from the field gun. Note the limited protection offered by this arrangement.

As a result of the field tests, a number of modifications were made to the T12. A new shield was required offering better protection for the gun crew. Several concepts were considered before the adoption of a low silhouette shield sloped on the front and top. Originally, the T12 was manned by a crew of four consisting of the driver, assistant driver (radio operator), gunner, and loader. The Armored Force Board reported that this was inadequate and that two additional seats for crew members should be provided. Eventually, the crew was set at five men including a gun commander. The .50 caliber machine gun and its ammunition were eliminated and two sheet metal stowage boxes were added to the rear of the vehicle. The standard fire control equipment for the 75mm field gun originally proposed was not available and the Ordnance Committee authorized the use of the telescope M33, telescope mount M36, and instrument light M17. Although the T12 had been standardized as the 75mm gun motor carriage M3 in October 1941, the later modifications were applied to the new vehicles then in production. Also, the changes specified for the later production M3 personnel carriers were applied to the gun motor carriages.

The three photographs at the right show the half-track 75mm gun motor carriage T12 with an interim shield design which, although it was an improvement, was not considered to be satisfactory.

The final version of the gun shield is shown on the half-track 75mm gun motor carriage T12 at Aberdeen. The upper and lower photographs were dated 29 November 1941 and 4 April 1942 respectively. A pedestal mount for a .50 caliber M2HB machine gun has been installed on the vehicle in the lower photograph.

The improved protection offered by the final shield design can be seen in the photograph above. Additional details of the half-track 75mm gun motor carriage, now designated as the M3, can be seen in the views at the right and below during its evaluation by the troops.

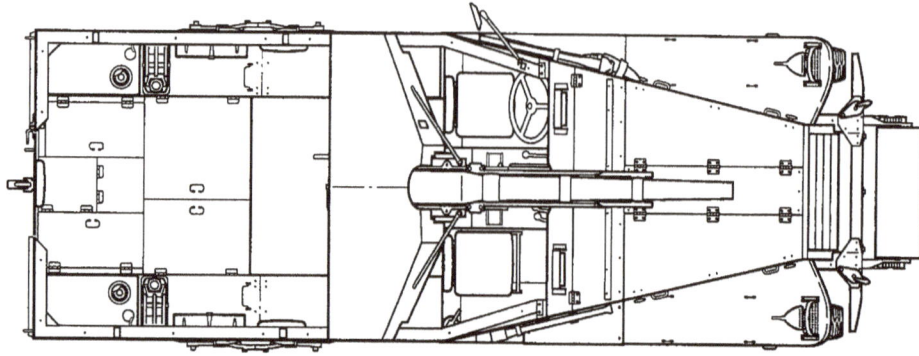

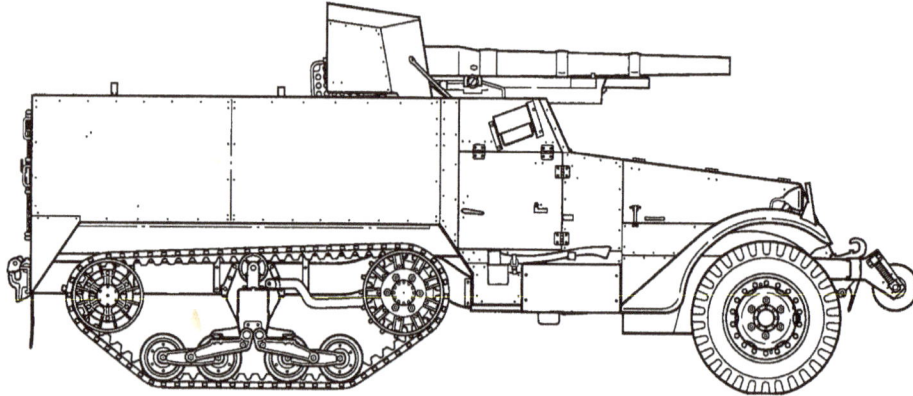

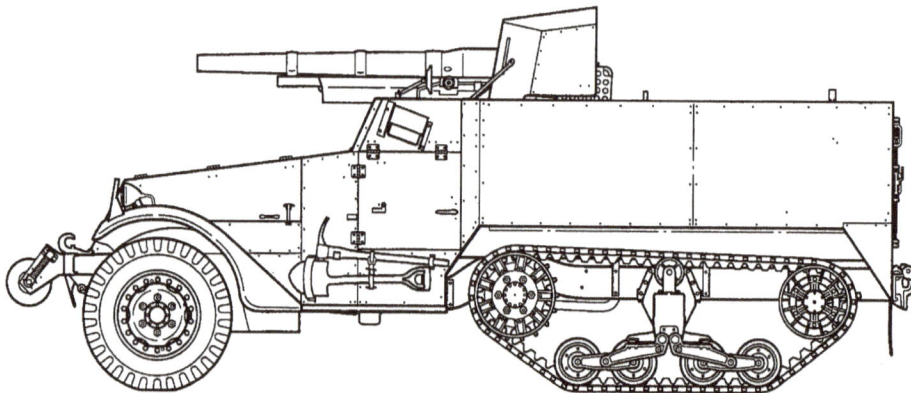

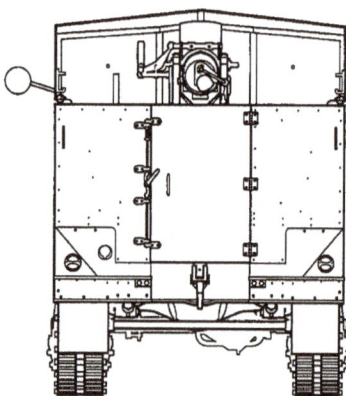

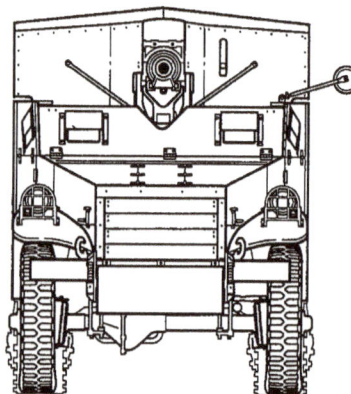

Scale 1:48

Half-Track 75mm Gun Motor Carriage M3

Above, the half-track 75mm gun motor carriage M3 is shown with the stowage compartments closed (left) and open (right). Note that the vehicle at the left has the demountable headlights and the one at the right retains the early fixed headlights. Below, the M3 has the early headlights. The two views at the right show the demountable headlights and the double coil spring on the idler. The letters in the photograph at the right indicate various segments of armor plate.

The half-track 75mm gun motor carriage M3 above has been modified by the installation of an indirect sighting device on the top of the shield. This photograph was taken at Aberdeen Proving Ground on 20 October 1942. Below at the right, the shield has been removed from this M3 and a T18 computing sight has been installed.

The requirement for new 75mm gun motor carriages exceeded the supply of M2A3 gun carriages needed for conversion to the M3 mount. However, quantities of the earlier M2A2 75mm field gun carriage were available and parts from this mount were adapted for installation in the gun motor carriage. In July 1942, Ordnance Committee action designated the mount using parts of the M2A2 field carriage as the 75mm gun mount M5 and the vehicle armed with that mount as the 75mm gun motor carriage M3A1. Tests at Aberdeen concluded that the M5 was a satisfactory mount. The traverse range increased to 42 degrees (21 degrees right and 21 degrees left). Maximum elevation remained at 29 degrees, but the maximum depression was less than 7 degrees.

Production of the 75mm gun motor carriage M3 and M3A1 ran from February 1942 until April 1943 for a total of 2116 vehicles in addition to the original 86 T12s.

The 75mm gun mount M5 appears at the right installed in the half-track 75mm gun motor carriage M3A1. The mount shown here was at Aberdeen Proving Ground on 22 January 1943.

The half-track 75mm gun motor carriage T73 appears in the photograph above at Aberdeen Proving Ground on 2 September 1943. Below at the right is a rear view of the 75mm gun T15 in mount T17 as used in the 75mm gun motor carriage T73.

In early 1943, the supply of 75mm guns M1897A4 was about exhausted. At that time, the M3 gun motor carriage was expected to continue in production. In March, authority was given to modify the 75mm gun M3, then used in the Sherman tank, for installation in the gun motor carriage. A sleigh was designed and the 75mm gun M3 was reworked to permit its installation in the M2 75mm gun recoil mechanism. The modified weapon was designated as the 75mm gun T15 in the mount T17. When installed in the half-track, the vehicle became the 75mm gun motor carriage T73.

Tests of the T73 at Aberdeen Proving Ground indicated that the new gun motor carriage was satisfactory after minor modifications. However, by this time, greatly improved tank destroyers were available and the requirement for the expedient vehicle no longer existed. The T73 project was canceled and 1360 of the M3 and M3A1 gun motor carriages were converted to the half-track personnel carrier M3A1. The 75mm gun motor carriages M3 and M3A1 were classified as Limited Standard in March 1944 and declared obsolete in August of the same year.

At the right is a line up of half-track 75mm gun motor carriages M3 during the training of the early tank destroyer units.

Above, the pilot half-track gun motor carriage T48 is at Aberdeen Proving Ground on 16 June 1942. The vehicle is armed with the short barrel Mark III 6 pounder (57mm).

The successful employment of the 75mm gun motor carriage M3 led to the adaptation of the half-track as the self-propelled mount for other heavy weapons. On 15 April 1942, OCM 18099 recommended the development of the 57mm gun motor carriage T48 based upon the half-track personnel carrier M3.

Intended primarily for Great Britain under the Lend-Lease program, the T48 was armed with the British 6 pounder antitank gun. Two versions of this weapon mounted in combat vehicles were designated as the ordnance, quick firing, 6 pounder Mark III and Mark V. The Mark V was manufactured in the United States as the 57mm gun M1. The two weapons differed only in the barrel. The Mark III had a heavy wall tube with a bore length of 42.9 calibers compared to the thin wall tube 50 calibers in length on the Mark V. The thick wall tube on the Mark III made it heavier than the Mark V.

Assembly of the pilot 57mm gun motor carriage T48 was in progress by May 1942 at Aberdeen Proving Ground. The 57mm gun was mounted in the M12 recoil

mechanism and installed on a tubular pedestal using the top portion of the M1 field carriage for the 57mm gun. To simplify manufacturing, the original pedestal was replaced by a conical structure and it was designated as the 57mm gun mount T5. The gun was aimed forward and mounted just behind the driving compartment on the half-track personnel carrier M3. On this mount, it had a total traverse of 55 degrees ($27^1/_2$ degrees left, $27^1/_2$ degrees right) and the elevation ranged from +15 to -5 degrees. The short barreled British Mark III 6 pounder was installed in the pilot, but the longer 57mm gun M1 was specified for the production vehicles. The original travel lock installed in the recoil mechanism of the mount proved to be unsatisfactory and it was replaced by a travel lock mounted on the vehicle hood which engaged the muzzle of the gun. The armor plate windshield cover was hinged at the bottom as on the 75mm gun motor carriage M3 and had a notch in the center to provide clearance for the gun tube. Initially, the gun shield from the wheeled 57mm gun motor

Below, the Mark III 6 pounder (57mm) gun is at maximum depression (left) and maximum elevation (right) when firing toward the front. Note the early shield design on the pilot.

Additional views of the half-track 57mm gun motor carriage T48 pilot show the vehicle with the original shield above and an improved design below. The latter photographs were dated 3 August 1942

carriage T44 was installed on the T48 to expedite the test program. After the first tests were complete, a new shield was designed using face hardened steel armor $^5/_8$ inches thick in front and $^1/_4$ inch thick on the sides and top. This shield extended back over the gun crew and provided the vehicle with a relatively low silhouette for a weapon mounted on a half-track. The height over the gun shield was only 90 inches. Because of the experience with the 75mm gun motor carriage M3, the new demountable headlights were installed to prevent damage from the muzzle blast of the gun. This muzzle blast produced considerable force and when the gun was at low elevation resulted in deformation of the vehicle

Although the pilot half-track 57mm gun motor carriage T48 is based upon an early production half-track with the fixed rear idler, It has been fitted with the demountable headlights to prevent damage from the muzzle blast of the 57mm gun.

Details of the half-track 57mm gun motor carriage T48 pilot can be seen in the photographs on this page. They were all dated 3 August 1942 at Aberdeen Proving Ground. Above, the driver's compartment appears at the left and the front of the gun mount can be seen at the right. A rear view of the gun mount is shown below at the right.

hood. This problem was eliminated by the addition of reinforcing angles to the hood plates. The T48 was manned by a crew of five men.

Production of the 57mm gun motor carriage T48 began at the Diamond T Motor Car Company in December 1942 and continued until May 1943. A total of 962 gun motor carriages were completed. Although the T48 was originally intended for Britain, only 30 were allocated and 650 were sent to the Soviet Union. One vehicle went to the U.S. Army overseas and the remaining 281 were converted to M3A1 half-track personnel carriers at the Chester Tank Depot.

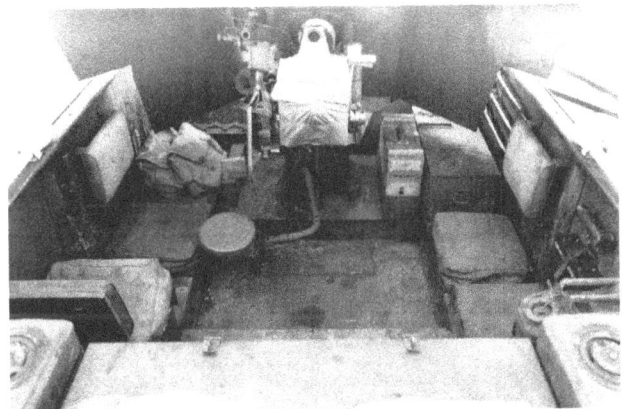

Below at the left, the ready round rack is open showing the 57mm ammunition and at the right, the ready round rack is closed and the lower ammunition rack is visible. Note the 57mm ammunition in their fiber containers. The fuel tanks are on each side of the ammunition racks.

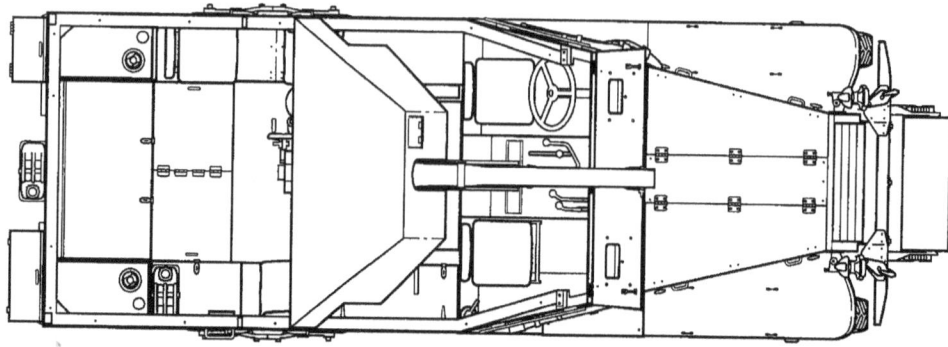

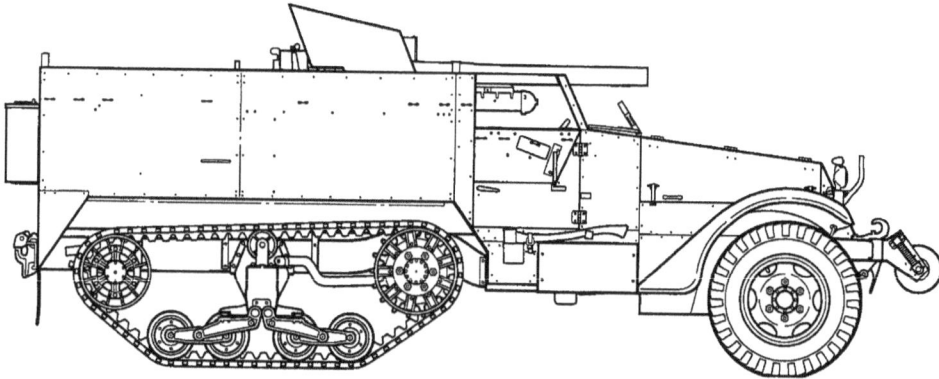

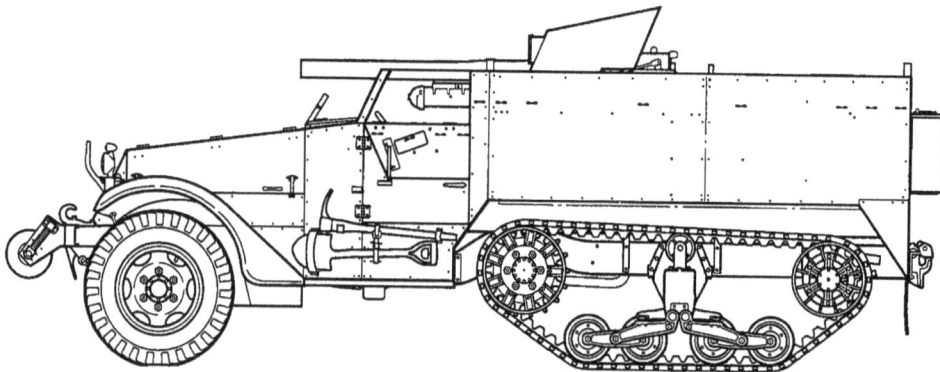

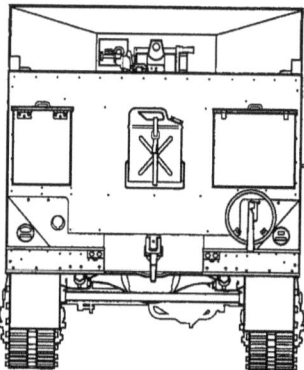

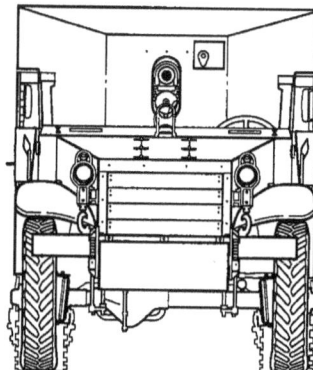

Half-Track 57mm Gun Motor Carriage T48, Pilot

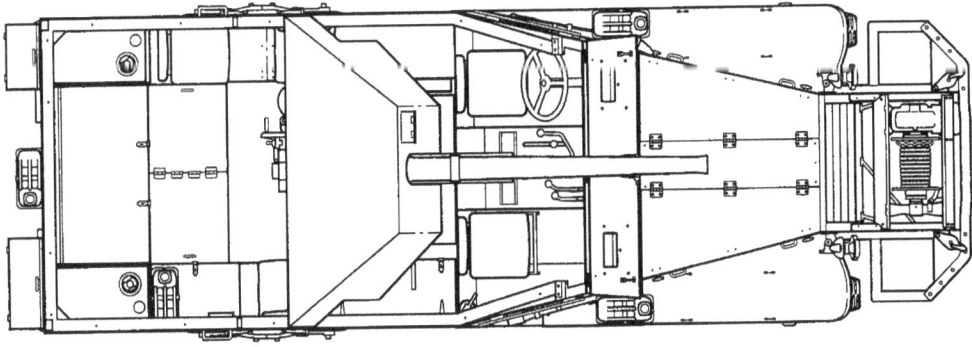

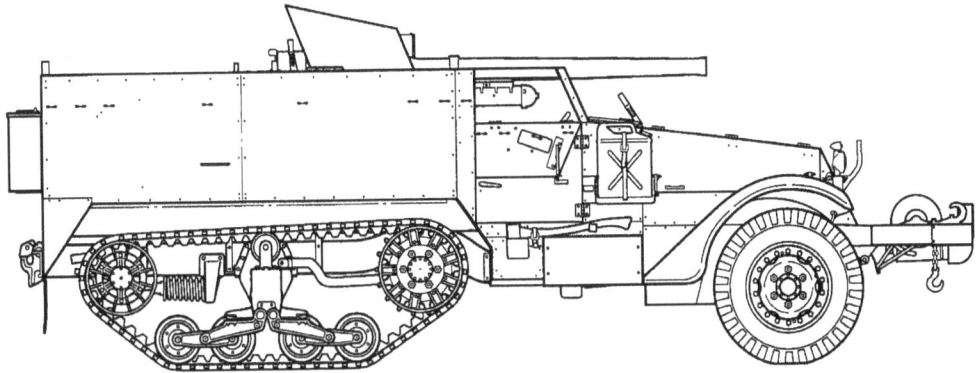

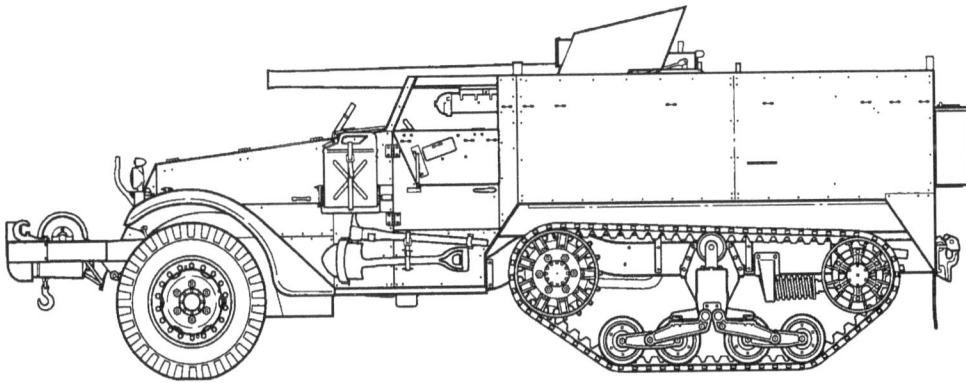

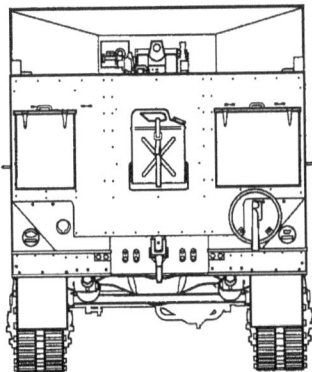

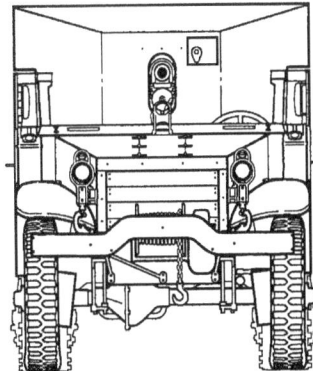

Scale 1:48

©M. Duplessis

Half-Track 57mm Gun Motor Carriage T48, production

The pilot half-track 105mm howitzer motor carriage T19 is shown in the photographs on this page. The pilot T19 was at Aberdeen Proving Ground on 29 November 1941.

SELF-PROPELLED ARTILLERY

As the Armored Force expanded during the Fall of 1941, there was an urgent need for self-propelled artillery. Although full track chassis were preferred, the situation required the use of whatever vehicles were immediately available. Once again, the job fell on the half-track personnel carrier M3 and it was selected to carry the 105mm howitzer M2A1. Although this combination had originally been suggested in September, it had been disapproved. However, the urgency of the requirement resulted in its approval by the Adjutant General and the construction of a pilot was authorized by OCM 17391, dated 31 October 1941. This action designated the new vehicle as the 105mm howitzer motor carriage T19. As with the other early pilots, the T19 was assembled at Aberdeen Proving Ground using the recoil mechanism M2 and parts of the carriage M2. The test firing gave better results than

expected, but several runs over rough terrain caused the frame to sag. To correct this problem, the frame was reinforced and the howitzer mount was redesigned to spread the load over a greater area. Demountable

These views show the half-track 105mm howitzer motor carriage T19 pilot at Aberdeen. No shield has yet been installed on the howitzer.

headlights were recommended to reduce the damage from muzzle blast, but they were unavailable for the early vehicles. Initially the M2A1 105mm howitzer was mounted on the vehicle without a shield. After some experiments, a folding shield similar to that on the towed howitzer was installed. Like the guns fitted to the half-track chassis, it was mounted firing forward. The total traverse of the howitzer was 40 degrees (20 degrees left, 20 degrees right) and the elevation ranged from +35 to -5 degrees. The armored windshield cover was remounted and hinged at the bottom so that it folded forward onto the hood. A crew of six was specified, but this was often increased in service.

After further tests at Aberdeen, a pilot was shipped to the Diamond T Motor Car Company to serve as a guide for production. The first production vehicle was delivered in January 1942. Production continued until April 1942 for a total of 324 T19s. Although the T19 was intended to be an expedient howitzer motor carriage until better full track vehicles were available, it served in North Africa and a few continued in action through Sicily, Italy, and into southern France. OCM 28557, dated 26 July 1945, approved the classification of the T19 as obsolete. That same month, the contractor Bowen & McLaughlin converted 90 T19s to M3A1 half-track personnel carriers.

The front and rear of the pilot half-track 105mm howitzer motor carriage T19 are shown above on 29 November 1941. Because of damage from muzzle blast, the demountable headlights seen at the right replaced the standard headlights on the T19. This photograph was dated 29 May 1942. An early attempt to provide more protection for the howitzer crew resulted in the installation of the experimental shield mounted on the pilot T19 in the view below.

The final design of the shield for the half-track 105mm howitzer motor carriage T19 appears in the photograph above taken at Aberdeen Proving Ground on 23 March 1942. A rear view of an earlier shield design can be seen at the right. This mount was photographed on 27 November 1941. Additional photographs of the final shield design appear below on 28 May 1942

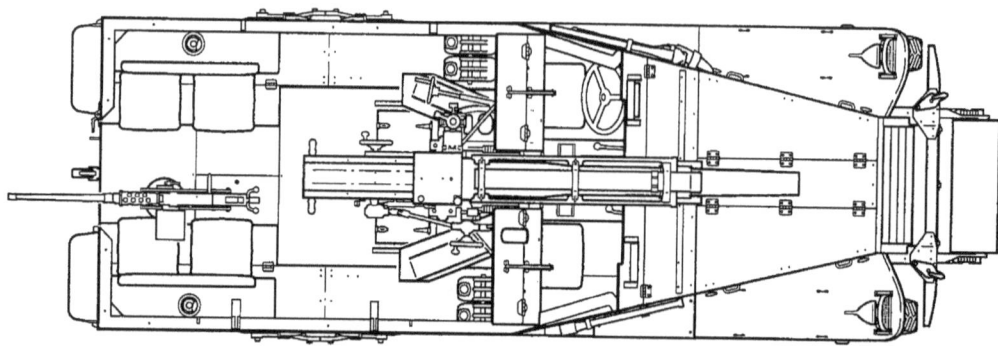

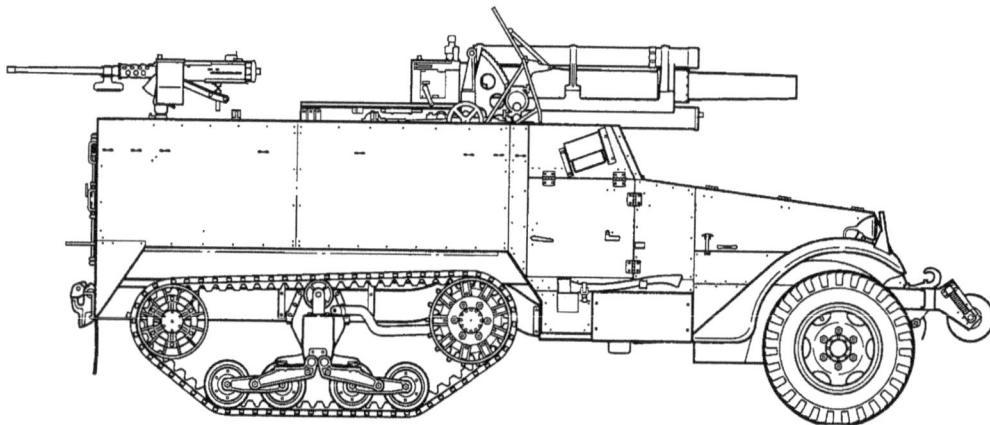

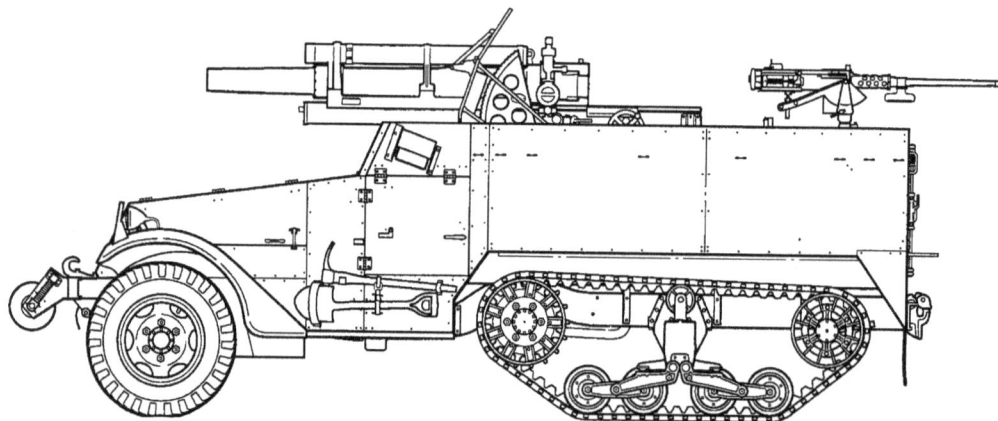

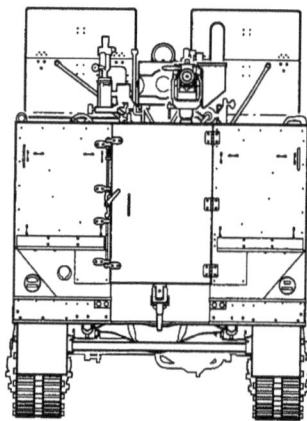

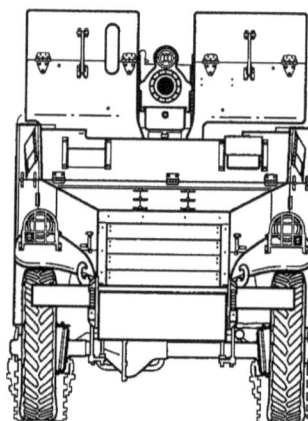

Scale 1:48

Half-Track 105mm Howitzer Motor Carriage T19

The early production half-track 105mm howitzer motor carriage T19 as shown here was not fitted with a shield on the howitzer. It also retains the early headlights.

The pilot half-track 75mm howitzer motor carriage T30 in these three photographs is at the Autocar Company on 22 January 1942. No shield is installed on the howitzer.

The success of the 75mm gun motor carriage M3 resulted in a request from the Armored Force Board for a similar installation of the 75mm howitzer M1A1 on the half-track personnel carrier M3. Authorized by verbal instructions from the Assistant Chief of Staff G4, a project was initiated for the self-propelled howitzer in October 1941. OCM 17665, dated 6 January 1942, designated the new vehicle as the 75mm howitzer motor carriage T30 and specified its characteristics. OCM 17809 approved this action which authorized the procurement of two pilot vehicles. The pilots were produced by the Autocar Company and one was shipped to Aberdeen Proving Ground for test and the other went to the White Motor Company to serve as a manufacturing pilot.

Initially, it was intended to pattern the shield after that on the M3A1 howitzer carriage, but on 23 March 1942, Ordnance requested that the height be reduced by at least six inches. Further modifications extended the shield eight inches to the rear and used a flap of armor

118

plate which moved with the howitzer to close the gap in the shield above the howitzer. In the final version of the T10 mount, the howitzer had a manual traverse of 45 degrees ($22\frac{1}{2}$ degrees left, $22\frac{1}{2}$ degrees right). The elevation range was from $+49\frac{1}{2}$ degrees to -9 degrees. The T30 carried a crew of five men.

Above, the howitzer in the half-track 75mm howitzer motor carriage T30 is at maximum elevation in this photograph dated 24 February 1942. Below, the howitzer is fitted with a revised shield on 23 April 1942. Both photographs were taken at Aberdeen Proving Ground.

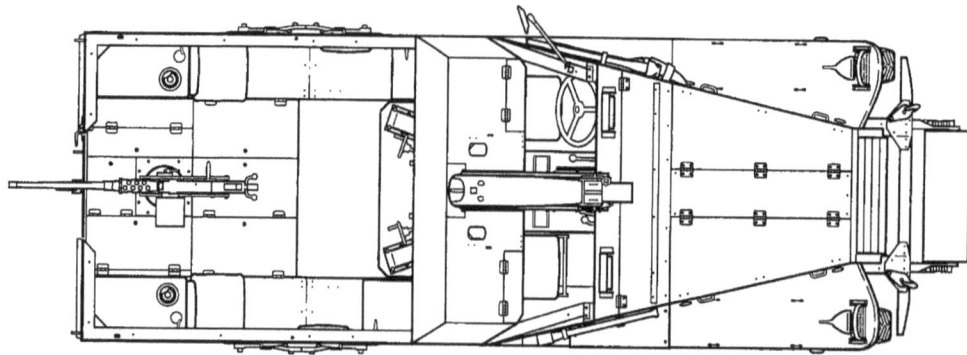

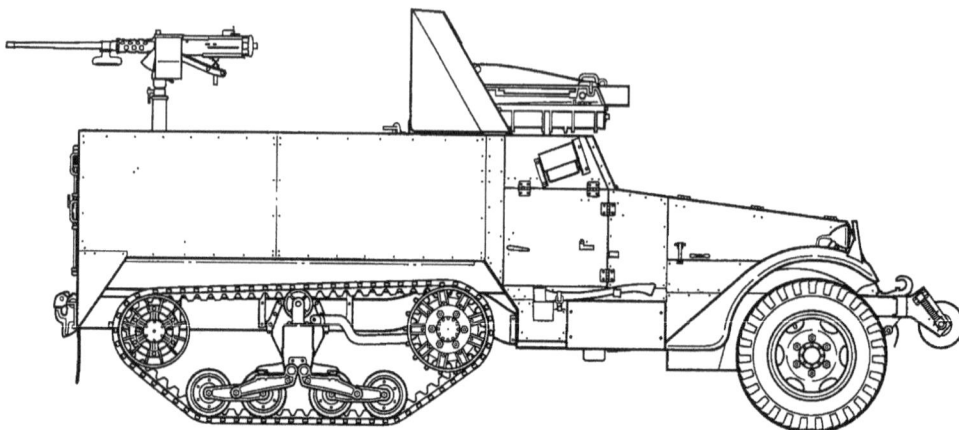

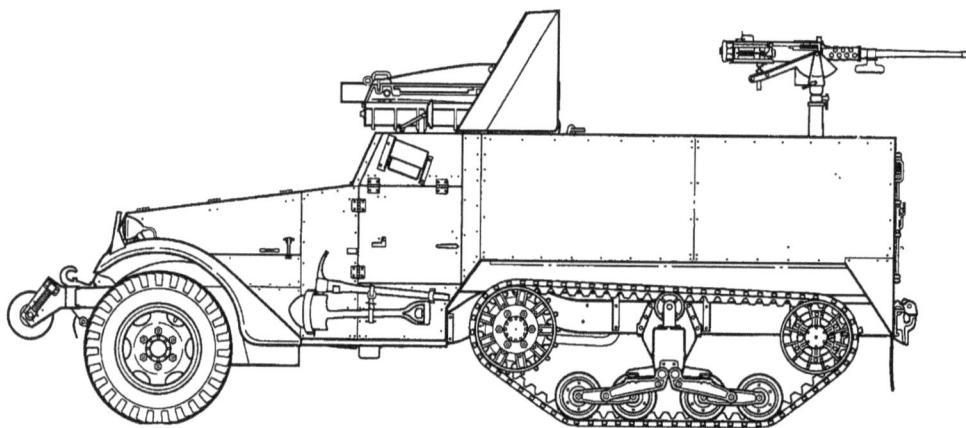

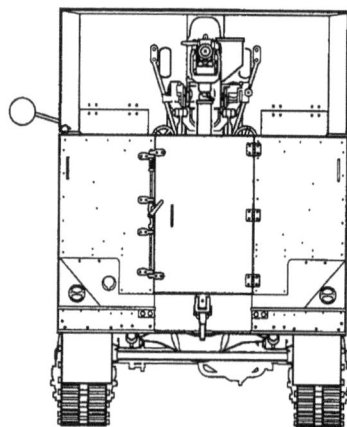

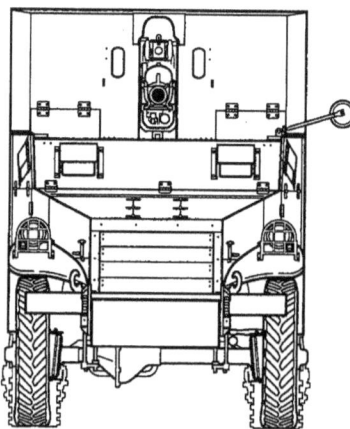

Scale 1:48

©M. Duplessis

Half-Track 75mm Howitzer Motor Carriage T30

Above and at the right are additional views of the half-track 75mm howitzer motor carriage T30 with the early design shield.

Production of the 75mm howitzer motor carriage T30 began at the White Motor Company in February 1942 and continued until April for a total of 312 vehicles. In August 1942, 108 of these were converted back to M3 half-track personnel carriers. In November 1942 an additional 188 T30s were delivered bringing the total manufactured to 500.

A design study considered the installation of the lightweight 105mm howitzer M3 on the half-track personnel carrier M3. This vehicle was designated as the 105mm howitzer motor carriage T38. However, the project was terminated and it was never built.

The proposed half-track 105mm howitzer motor carriage T38 is depicted in the side view drawing below.

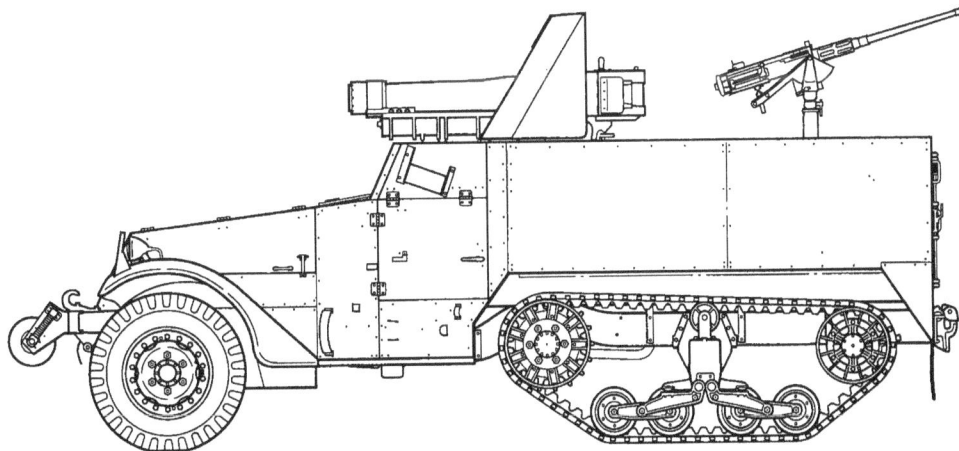

Scale 1:48

©M. Duplessis

Half-Track 105mm Howitzer Motor Carriage T38

The photographs above and below show the half-track multiple gun motor carriage T1E1 at Aberdeen Proving Ground on 22 August 1941.

ANTIAIRCRAFT VEHICLES

Heavy losses of vehicles from attacking aircraft during the early days of the war in Poland and France emphasized the need for antiaircraft protection of mobile columns. Several multiple machine gun mounts installed on trucks had been developed and tested at Aberdeen Proving Ground prior to World War II. Taking advantage of that experience, it was proposed to install some of the newly developed aircraft machine gun turrets on wheeled vehicles to provide a highly mobile antiaircraft weapon.

In October 1940, the Ordnance Committee recommended that a pilot multiple gun motor carriage T1 be developed utilizing the latest type of aircraft multiple gun mounts. These mounts were required to be power operated with a traverse of 360 degrees at a rate of 60 degrees per second and an elevation range of +80 to -5 degrees at a rate of 30 degrees per second. To meet this requirement, a twin .50 caliber machine gun turret was procured from the Bendix Aviation Corporation. This turret had an elevation range of +88 to -6 degrees and the maximum rate of traverse and elevation was 45 degrees per second. Power was supplied to the turret by a 30 volt motor generator set and a 24 volt battery.

The Bendix turret was tested at Aberdeen during June and July 1941 on a wooden test mount and a $\frac{1}{2}$ ton 4x4 truck. These tests showed excessive dispersion, particularly when firing toward the side on the truck mount. In an effort to find a more stable mount, it was recommended that the turret be installed on the $1\frac{1}{2}$ ton

122

Above, the Maxson twin machine gun mount (M33) appears at the left installed on the half-track multiple gun motor carriage T1E2. The complete vehicle appears at the right. Note that the fuel tank has been relocated to just behind the driver's compartment.

truck and on a half-track vehicle. An M2 half-track car was provided and the new antiaircraft vehicle was designated as the multiple gun motor carriage T1E1. However, problems with the turret required its return to the manufacturer for repair and during its absence a new turret appeared on the scene. This was the turret developed by the W. L. Maxson Corporation. The success of this turret resulted in the cancellation of the T1E1 program in April 1942.

A model of a twin .50 caliber machine gun turret was submitted by Maxson in November 1941 and a pilot was authorized for installation on a half-track car M2. The combination was designated as the multiple gun motor carriage T1E2. The pilot turret was completed in early 1942 and evaluated at Aberdeen Proving Ground on both the half-track and a fixed mount. The new experimental turret had an elevation range of +90 to -$11^1/_2$ degrees and it could be elevated and traversed at a maximum rate of 72 and 74 degrees per second respectively. It also was operated using its self-contained batteries and motor generator set. Using a V-belt drive controlled by simple linkages, the Maxson turret was rugged and less complicated than other turrets available and could be easily repaired by mechanics in the field. As a result of the favorable test report from Aberdeen, the new turret was standardized in March 1942 as the twin .50 caliber machine gun mount M33. In addition to the half-track, the M33 mount was installed for test purposes on the 4x4 $^3/_4$ ton truck, the 6x6 $1^1/_2$ ton truck, and the 6x6 $2^1/_2$ ton truck. As might be expected, the half-track provided the most stable mount. However, the half-track personnel carrier M3 was selected as the basis for the new antiaircraft vehicle because of the extra space provided by its longer body. The pilot of the new vehicle was designated as the multiple gun motor carriage T1E4.

Below are two views of the half-track multiple gun motor carriage T1E4 at Aberdeen Proving Ground on 8 July 1942. Note the folding armor panels at the top on the sides and rear to increase the maximum depression for the machine guns.

The half-track multiple gun motor carriage M13 appears above. The armored shield for the gunner has not been installed on this vehicle. In the view at the right, the armored shield is on the mount and all three photographs show the side and rear armor folded down in the firing position.

After some changes to the military characteristics, the standardization of the T1E4 was approved in September 1942 and it was designated as the multiple gun motor carriage M13. The first production pilot arrived at Aberdeen Proving Ground in December 1942. The upper part of the side and rear vertical armor was hinged to permit firing at -10 degrees elevation. Over the front of the vehicle, the guns could not be depressed below +30 degrees. A total of 1103 M13s were produced and all were completed at the White Motor Company from January through May 1943. Only 139 M13s were sent overseas and most of the remainder were converted to the later multiple gun motor carriage M16.

Below, the twin .50 caliber machine gun mount M33 is shown with the gunner's armor shield removed. The armor has been installed on the mount in the M13 at the right.

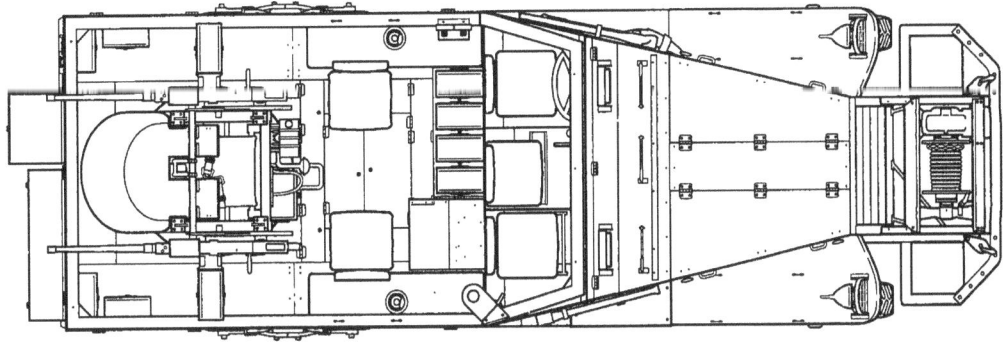

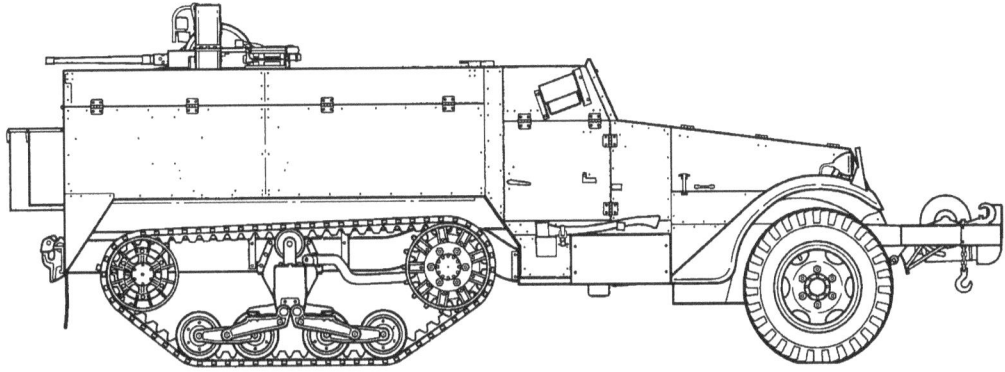

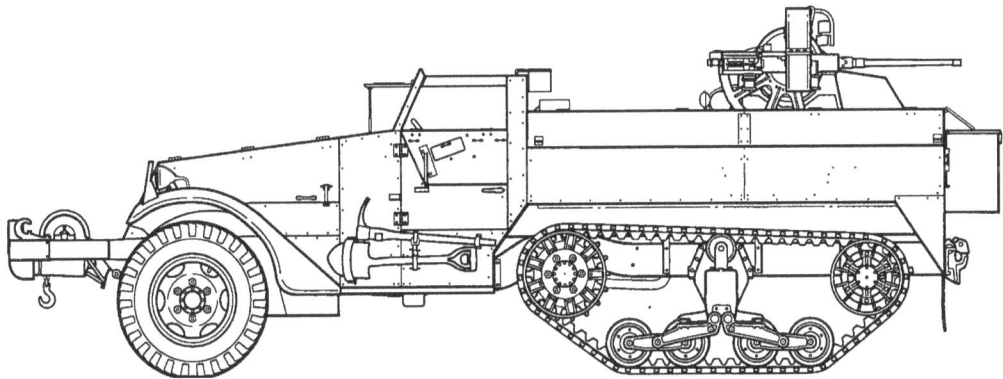

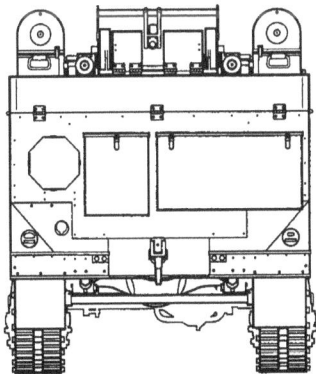

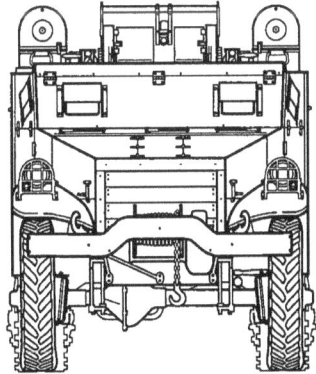

Scale 1:48

Half-Track Multiple Gun Motor Carriage M13

Details of the half-track multiple gun motor carriage M14 can be seen in the photographs on this page. Note that the armor shield for the gunner has not been fitted to the mount.

RA PD 18534

The M33 twin .50 caliber machine gun mount also was installed on the M5 half-track personnel carrier. Designated as the multiple gun motor carriage M14, these vehicles were intended for foreign aid under the Lend-Lease program. A total of 1605 M14s were manufactured by International Harvester during a production run extending from December 1942 through December 1943. Later, most of these had the gun mount removed and were used as personnel carriers in the British and Canadian Armies.

The gun mount on the M14 at the right is still without the armor shield. The two views of the M14 below show the gun mount armor installed. These two photographs were taken at Aberdeen Proving Ground on 26 October 1943.

126

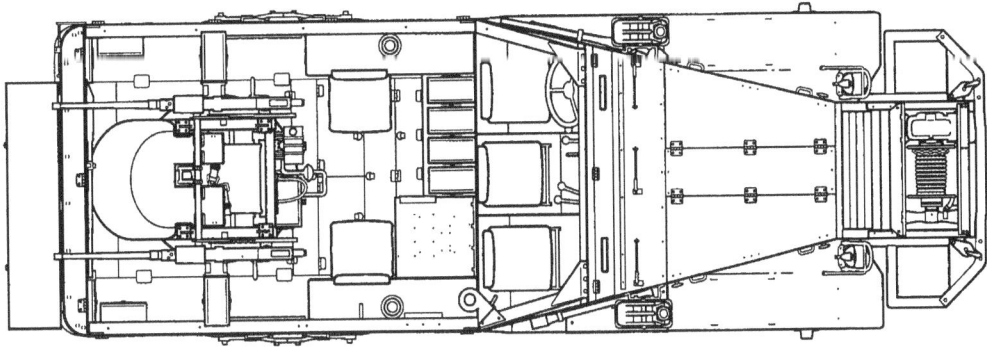

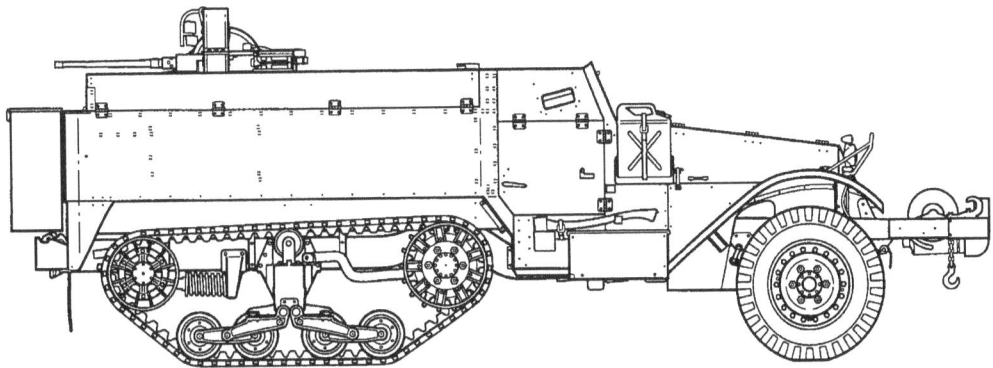

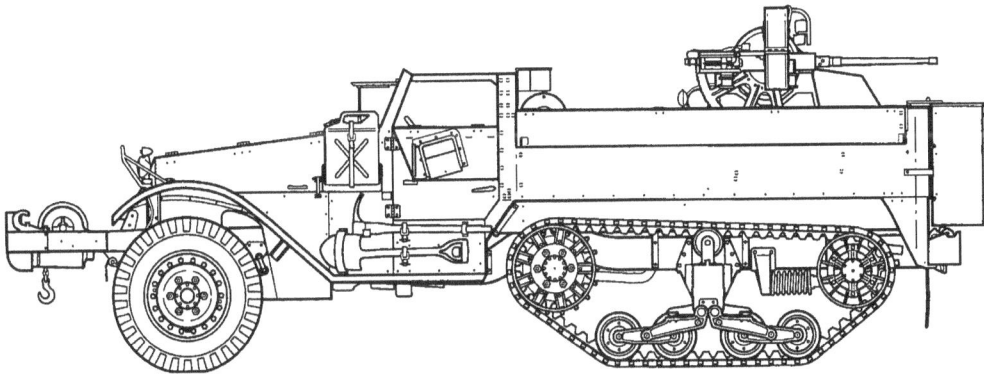

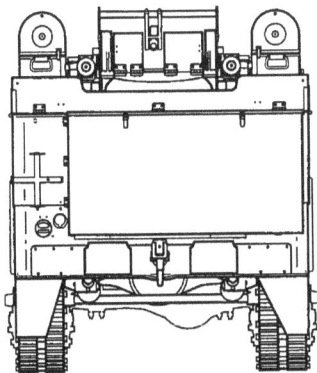

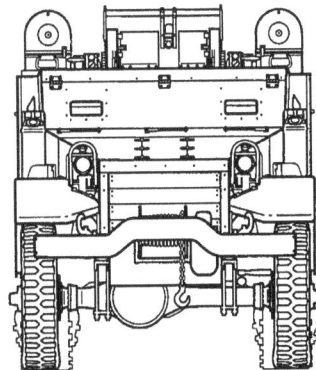

Scale 1:48

©M. Duplessis

Half-Track Multiple Gun Motor Carriage M14

Above, the proposed half-track multiple gun motor carriage T1E3 is illustrated at the left and at the right, the Martin twin .50 caliber machine gun turret is installed on the half-track of the multiple gun motor carriage T60E1 for a road test. The latter photograph taken at Aberdeen Proving Ground was dated 25 January 1944.

Several different aircraft gun turrets were considered prior to the selection of the Maxson mount. Among these was an electro-dynamic power turret developed by the Ordnance Section at Wright Field, Ohio. Procurement of one of these twin .50 caliber machine gun turrets was authorized in December 1941 for comparison testing along with the Bendix and Maxson turrets. The Ordnance Committee intended to mount this turret on the half-track car M2 and designated the vehicle as the multiple gun motor carriage T1E3. A standard Martin twin .50 caliber machine gun aircraft turret was modified to use the electro-dynamic drive and evaluated at Aberdeen on a $1\frac{1}{2}$ ton Air Corps trainer truck instead of the half-track. However, a Martin turret was installed later on the half-track chassis from the multiple gun motor carriage T60E1. Although the performance was inferior to the later version of the Maxson turret already standardized, the test program was continued to evaluate the technology for possible use in future turret development. However, the test results were generally unsatisfactory and the T1E3 program was closed in May 1944.

To comply with a request from the Chief of Coast Artillery, four experimental multiple gun motor carriages were built during September 1941. These vehicles were shipped to Aberdeen Proving Ground for tests during October and OCM 17313, dated 9 October 1941, assigned the designation multiple gun motor carriage T28. The new vehicle utilized the chassis of the half-track car M2 with the side and rear armor removed. The armament consisted of one 37mm gun M1A2 and two M2, water cooled, .50 caliber machine guns which were installed on the top part of the 37mm gun carriage M3E1. The four T28 pilots were as follows:

Serial No. 1875 with rotating platform
Serial No. 1862 with rotating platform and shield
Serial No. 1858 without rotating platform
Serial No. 1874 without rotating platform

Above, the 37mm automatic gun M1A2 is at the right. Below are two of the pilot half-track multiple gun motor carriages T28 before the twin .50 caliber machine guns were installed. Ordnance serial number 1875 is at the left and serial number 1862 is at the right. Note the locations of the fuel tank on these vehicles.

The half-track multiple gun motor carriage T28, serial number 1875, is shown on this page after installation of the two .50 caliber M2, water cooled, machine guns. These photographs from Aberdeen Proving Ground were dated 7 October 1941. Note the limited depression of the guns when firing forward over the driver's compartment.

The tests at Aberdeen indicated satisfactory performance in regard to the stability and dispersion of the motor carriage and the tracking characteristics of the mount. However, because of the limited space on the M2 chassis, it was recommended that the gun mount be installed on the larger half-track personnel carrier M3. Unfortunately, the Coast Artillery Board did not agree with these conclusions. According to their tests, the vehicle was relatively unstable and, as a consequence, inaccurate. They recommended the development of a new multiple gun motor carriage armed with four .50 caliber machine guns. As a result, the T28 project was closed in April 1942.

The half-track multiple gun motor carriage T28E1 number 1 appears above and below at Aberdeen Proving Ground on 31 August 1942. Note the changes in the gun mount compared to the T28. The .50 caliber ammunition stowage can be seen at the rear of the vehicle.

During this period, American observers with the British Eighth Army in the desert had noted a requirement for dual purpose antitank-antiaircraft weapons. Major General Charles L. Scott, former commander of the 2nd Armored Division, recommended the manufacture of such dual purpose self-propelled weapons armed with a 37mm gun and .50 caliber machine guns. In June 1942, a memorandum from the Services of Supply directed the procurement of 80 improvised 37mm self-propelled guns suitable for employment against both ground and aerial targets for a special mission. To Ordnance this looked like a job for a modified T28. On 9 July 1942, OCM 18477 directed the manufacture of the 80 vehicles based upon the half-track personnel carrier M3 and designated the new vehicle as the multiple gun motor carriage T28E1. Production of the T28E1 was at the Autocar Company during July and August 1942. In September, 78 of the new vehicles were assigned to the 443rd Antiaircraft Artillery, Automatic Weapons Battalion, Self-propelled. Two of the T28E1s were kept for experimental purposes. At that time, the 443rd was organized into four batteries with 20 guns each. Since they were short two guns, Battery D received two towed 40mm guns to bring them up to full strength.

As on the T28, the half-track personnel carrier M3 was modified by removing the side and rear armor and relocating the fuel tanks. The top part of the M3E1 37mm gun carriage was installed mounting the automatic 37mm gun M1A2 along with two M2 .50 caliber, water-cooled, machine guns. The weapons had a

360 degree traverse and the elevation ranged from 0 to +80 degrees except over the front where the guns could not be depressed below +20 degrees. Stowage was provided for 140 rounds of 37mm ammunition and 3400 .50 caliber rounds. Ten ammunition chests filled with 2000 .50 caliber rounds were ready for immediate use and the remaining 1400 rounds were stowed in a compartment at the rear of the vehicle.

When the 443rd received the T28E1s they had to make a number of modifications. When the elevation was decreased to +15 degrees, the 37mm ammunition clip would eject against the frame and cause a jam. The mechanics in the 443rd modified the carriage not only correcting the problem, but increasing the range of elevation down to -5 degrees. At that time, Army radios were in short supply so the T28E1s were fitted with civilian radios. Thus they left for war and the invasion of North Africa.

The .50 caliber M2 water cooled machine gun is at the right.

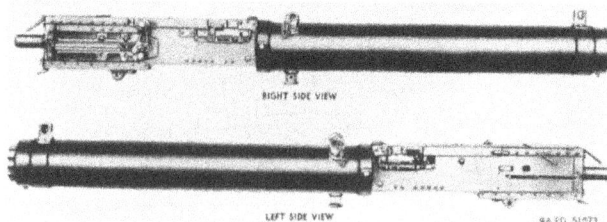

Above, the production pilot of the half-track multiple gun motor carriage M15 is at the Autocar Company on 8 December 1942.

One of the original T28 vehicles was evaluated along with the 40mm gun motor carriage T54 by the Antiaircraft Artillery Board at Camp Davis, North Carolina. Neither was considered satisfactory, but the T28 was more advanced than the T54. The Board concluded that the T28 would be useful as an expedient weapon if a number of modifications were made. These included the use of the M3 half-track personnel carrier as had already been done on the T28E1. Also, the water-cooled machine guns were to be replaced by air-cooled, heavy barrel, .50 caliber weapons with an improved firing mechanism. Other changes reversed the positions of the lead setter's hand wheels, shifted the vertical hand wheel into the fore and aft plane, installed larger hand wheels, provided seats for two lead setters in place of one, installed an armored shield, improved the bore sighting arrangement, provided a means for the collection and disposal of empty cartridge cases, and extended the elevation range to -5 degrees.

On 29 October 1942, OCM 19087 designated the modified vehicle as the multiple gun motor carriage M15 and recommended its classification as Substitute Standard. The mount was designated as the combination gun mount M42. However, the maximum depression was retained at 0 degrees because of delays that would be caused by any change. The recommendation was approved in November 1942 and production was authorized for 600 vehicles to meet the requirement for expanding operations in North Africa. In December, the sighting system M6 was approved replacing the M2E1 system on the T28E1.

A production pilot M15 was built by the Autocar Company and evaluated at Aberdeen Proving Ground during December 1942 and January 1943. During this period, the movable shield was removed from the gun mount to improve the visibility. The 37mm ammunition stowage on the M15 was increased to 240 rounds, but the .50 caliber ammunition remained at 3400 rounds as on the T28E1. Production of the M15 began at Autocar in February 1943 and continued until April for the total run of 600 vehicles. A total of 289 T28E1 and M15 multiple gun motor carriages were deployed with the American troops overseas. They served in North Africa, Sicily, Italy, the invasion of Southern France and on into Germany by the end of the war in Europe. In August 1945, the M15 and the mount M42 was classified as obsolete.

The production pilot half-track multiple gun motor carriage M15 is at the right. This photograph taken at Aberdeen Proving Ground was dated 16 December 1942.

These are additional views of the production pilot half-track multiple gun motor carriage M15 at Aberdeen on 16 December 1942. Apparently, the armor plate on the front of the gun mount itself obstructed vision and it was soon discarded.

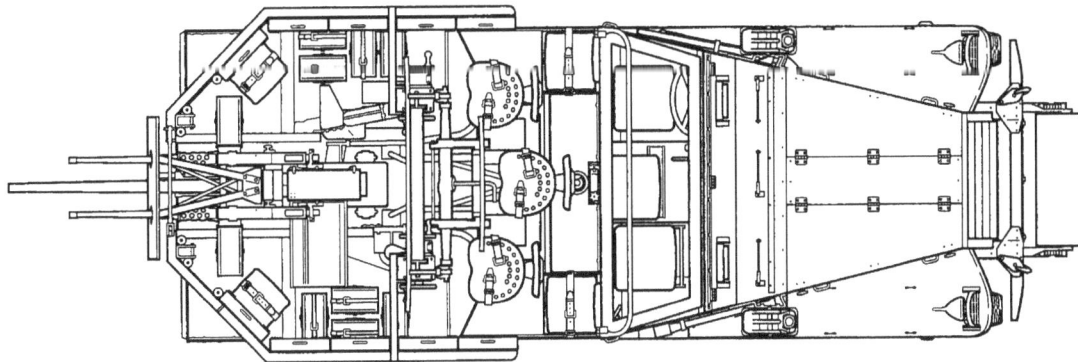

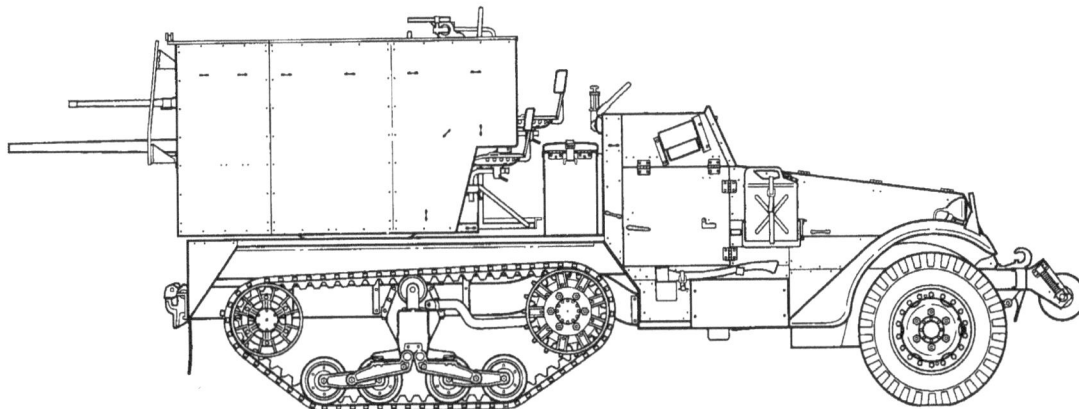

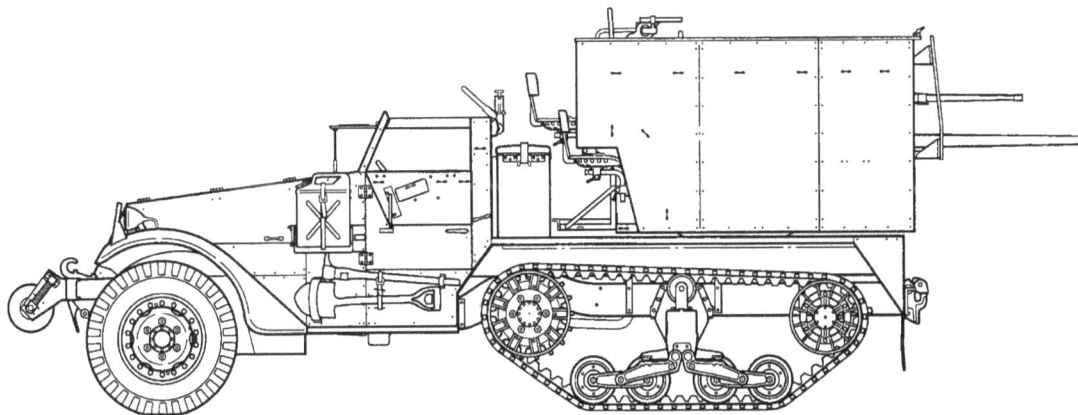

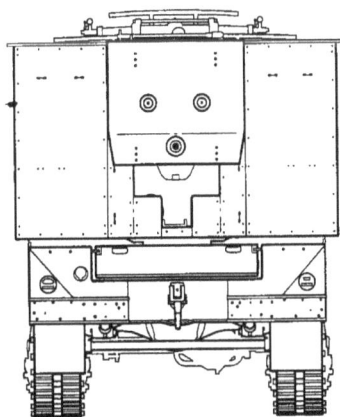

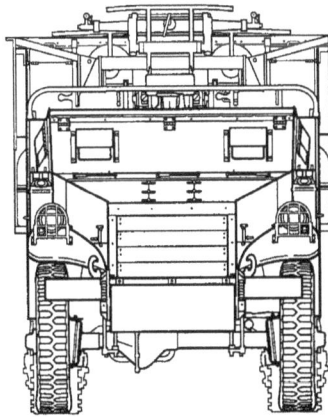

©M. Duplessis

Half-Track Multiple Gun Motor Carriage M15

After initial tests at Aberdeen, the M15 production pilot received several modifications. Among these was the elimination of the armor plate on the front of the gun mount. The modified vehicle is shown here on 13 January 1943.

The half-track combination gun motor carriage M15E1 above is at Aberdeen Proving Ground on 11 June 1943. At the bottom of the page are two views of the half-track combination gun motor carriage M15A1. Note the obvious changes compared to the M15. The .50 caliber machine guns are mounted below the 37mm gun and the top front armor on the turret folds down.

The successful employment of the multiple gun motor carriages during the North African campaign resulted in a requirement for additional vehicles. However, the supply of top carriages from the M3E1 37mm gun carriage which served as the basis for the M42 mount was exhausted. About 1750 surplus 37mm M3A1 gun carriages were available and authority was received to adapt them for the new multiple gun motor carriage. An M15 was modified by removing the M42 gun mount

and replacing it with the top carriage from the M3A1 gun carriage minus the remote control system. Two .50 caliber M2, heavy barrel, air-cooled machine guns and the sighting system M6 were installed on the new mount. OCM 21226, dated 5 August 1943, designated the modified vehicle as the combination gun motor carriage M15E1 with the combination gun mount T87. Note the change in nomenclature. After minor modifications, standardization was recommended by OCM

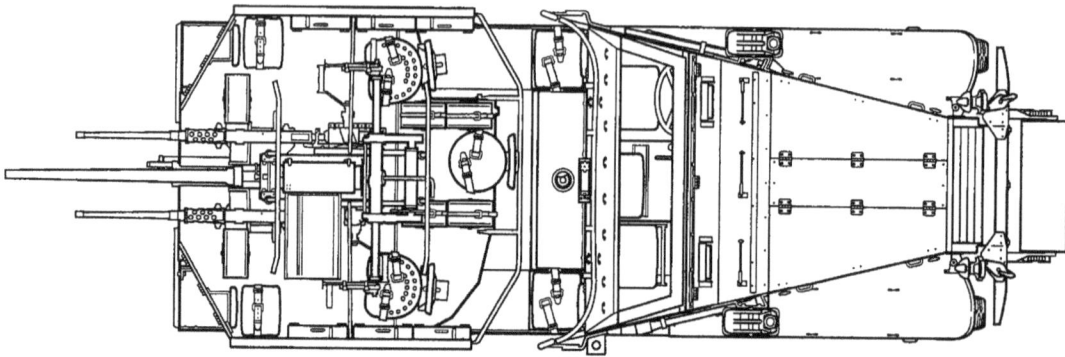

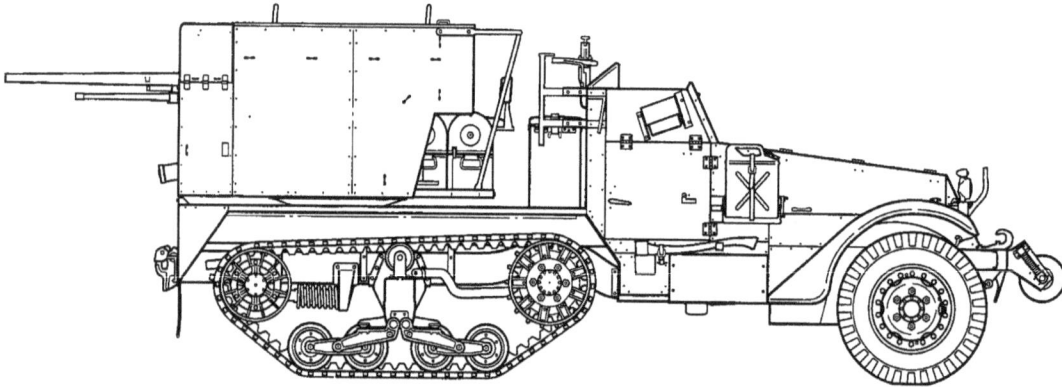

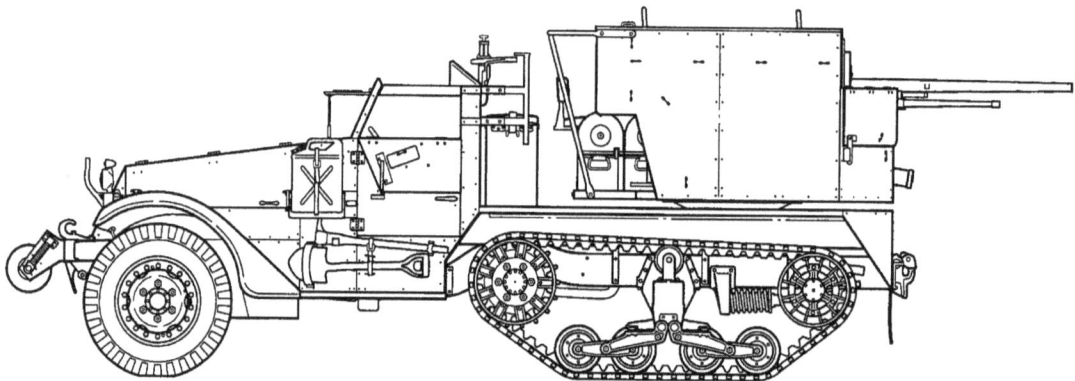

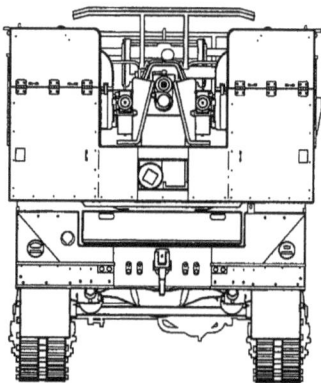

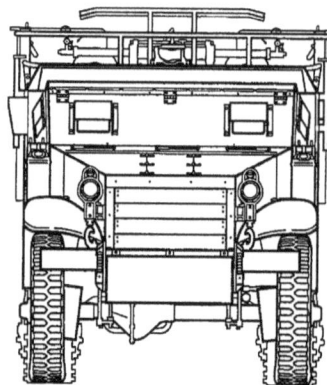

©M. Duplessis

Half-Track Combination Gun Motor Carriage M15A1

The half-track combination gun motor carriage M15A1 appears above and below at the right. The top front armor on the turret has been folded down above. Details of the .50 caliber M2HB machine gun are shown at the lower right.

21281 dated 12 August 1943. This action was approved in September. The vehicle was now designated as the combination gun motor carriage M15A1 with the combination gun mount M54. There were several obvious differences on the M15A1 compared to the earlier M15. On the new gun mount, the .50 caliber machine guns were mounted below the 37mm gun and the top panel of the front armor could be folded down on each side of the guns to improve visibility. Platforms were provided for the 37mm gun loader and the lead setter and the lead setter's seat was raised to allow operation of the sight from the seated position. A rail was added to the rear of the mount to aid the crew when mounting or dismounting. The ammunition stowage on the M15A1 was reduced to 200 37mm rounds and 1200 .50 caliber rounds. Production of the M15A1 began at the Autocar Company in October 1943 and continued until February 1944. A total of 1652 M15A1s were produced. OCM 22985, dated 24 February 1944, recommended standardization of the computing sight M14 for installation on the M15A1. This action was approved by OCM 23270. On 12 July 1945, OCM 28349 recommended the reclassification of the M15A1 as Substitute Standard. This action was approved and in August it was classified as Limited Standard. However, the M15A1 continued to serve in the Army until the end of the Korean War. During World War II, 100 of the M15 series vehicles were furnished to the Soviet Union under the Lend-Lease program.

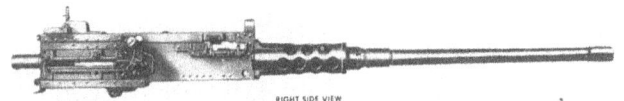

RIGHT SIDE VIEW

LEFT SIDE VIEW

The combination gun mount M54 can be seen at the right

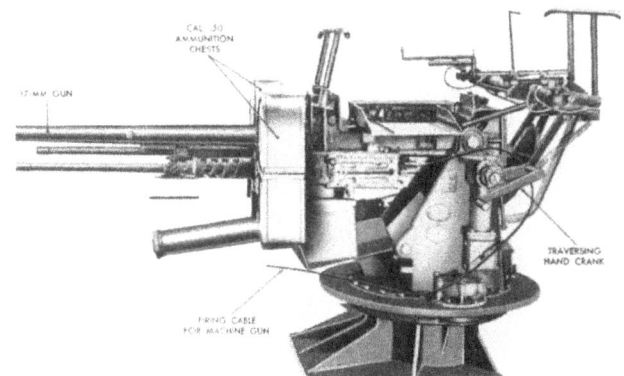

The arrangement of the guns on the 50 caliber multiple machine gun mounts T60 (left) and T60E1(right) can be seen above. Below are top views of the half-track multiple gun motor carriage T37 (left) and the half-track multiple gun motor carriage T37E1 (right). Note that both vehicles are armed with .50 caliber M2 aircraft machine guns.

As mentioned previously, after rejection of the multiple gun motor carriage T28 by the Coast Artillery Board, it was recommended that a preferable solution to the antiaircraft problem would be a new self-propelled mount armed with four .50 caliber machine guns. Approval for the development of such a self-propelled weapon was recorded in OCM 17548 dated 18

December 1941 and it was designated as the multiple gun motor carriage T37. Two mounts, each with four .50 caliber machine guns, were designed by the United Shoe Machinery Company. On 26 March 1942, OCM 17972 recommended procurement of two pilots for each type of mount. These vehicles were designated as the multiple gun motor carriage T37 with the machine gun mount T60 and the multiple gun motor carriage T37E1 with the machine gun mount T60E1. On the T60 mount, the machine guns were arranged in two rows one above the other with the two lower guns more closely spaced than the two upper guns. The T60E1 mount placed all four guns in a single row with the two outer guns extending far forward of the two inner guns. It is

The half-track multiple gun motor carriages T37 and T37E1 appear above at the left and right respectively. Note the frames attached to the front of each gun mount to limit the depression when firing forward over the driver's compartment. Below at the right is a left side view of the half-track multiple gun motor carriage T37E1. Both the T37 and T37E1 were based upon the early production half-track personnel carrier M3.

interesting to note that all of the machine guns were .50 caliber M2 aircraft type weapons with the lighter 36 inch barrel compared to the heavy 45 inch barrel used in most ground applications. Like the multiple gun motor carriage M15, each of the new mounts utilized the top portion of the 37mm gun carriage M3E1. Both mounts were mechanically operated requiring a three man crew with a separate man to elevate, traverse, and fire the weapons. The mount on both vehicles was protected by a $1/2$ inch thick circular armor shield. The weapons could be traversed 360 degrees and the elevation ranged from +85 to -5 degrees. After tests of the T37E1, the Antiaircraft Artillery Board concluded that it was unsatisfactory as an antiaircraft weapon. Also, since all available M3E1 gun carriages were being utilized in the manufacture of the multiple gun motor carriage M15, there was no future for the T37 or the T37E1. OCM 19511, dated 14 January 1943, records the closing of the project.

Below and above at the right are additional views of the half-track multiple gun motor carriage T37E1. These photographs were taken on 22 June 1942 at Aberdeen Proving Ground.

Above, the multiple .50 caliber machine gun mount T61 is installed on the half-track car M2, registration number 4011370. This was the same vehicle used for the pilot multiple gun motor carriage T1E2 and it retains those markings. At the bottom of the page is the pilot half-track multiple gun motor carriage M16. This was the vehicle originally designated as the multiple gun motor carriage T58. However, it was standardized as the M16 before it was completed. It is based upon the half-track personnel carrier M3.

Parallel with the work on the T37 and the T37E1 an effort was made to increase the firepower of the twin gun Maxson turret. In April 1942, development was authorized for the multiple .50 caliber machine gun mount T61. This was essentially a four gun version of the two gun Maxson mount standardized as the M33. The pilot T61 mount was delivered to Aberdeen Proving Ground in August 1942 and installed for test on the same M2 half-track car used to evaluate the two gun Maxson mount. The vehicle still bore the marking T1E2. However, the T61 was intended for eventual installation on the half-track personnel carrier M3 because of the greater space available. When mounted on the latter chassis, it was designated as the multiple gun motor carriage T58. Firing tests at Aberdeen and by the

Antiaircraft Artillery Board showed excellent results and with minor modifications, the T61 was recommended for standardization as the multiple .50 caliber machine gun mount M45 by OCM 19264 on 3 December 1942. The same action assigned the designation multiple gun motor carriage M16 to the vehicle based upon the half-track personnel carrier M3. If the M45 mount was installed upon the half-track personnel carrier M5, it became the multiple gun motor carriage M17. The latter vehicle was classified as Substitute Standard and was intended for foreign aid.

An M16 pilot vehicle was assembled at Aberdeen in early 1943 to work out minor modifications and stowage arrangements. A six inch adapter ring was bolted the bottom of the mount. This raised the level of

Additional views of the pilot half-track multiple gun motor carriage M16 (T58) are shown here. The vehicle was photographed at Aberdeen Proving Ground on 2 February 1943. Like most of the pilots, it made use of an early production half-track with the fixed idler and original headlights.

The production half-track multiple gun motor carriage M16 shown here is at the General Motors Proving Ground on 31 August 1943. Unlike the pilot, this M16 is fitted with a winch replacing the front roller and cutouts in the folding side and rear armor provide clearance for the ammunition chests when traversing the gun mount at low elevation. Note the demountable headlights.

the mount so that the guns could fire over the sides and rear when the armor flaps were folded down. The Armored Board at Fort Knox considered the vehicle satisfactory and recommended that it be placed in production immediately with minor changes. Production of the M16 began at the White Motor Company in May 1943, the same month the last of the M13 multiple gun motor carriages was delivered. Production of new M16s continued until March 1944 for a total of 2877. White also converted 568 M13s and 109 twin 20mm gun motor

carriages to the M16 configuration. An additional 60 M13s were converted to M16s by Diebold Incorporated bringing the total M16 production to 3614 vehicles. In September 1944, the Ordnance Committee recommended that the gun mounts be modified to add a rear platform to accommodate two cannoneers. When so modified the mount was designated as the M45D. The M16 was to have a long life in the U. S. Army. Subject to two postwar modifications, the series was not declared obsolete until February 1958.

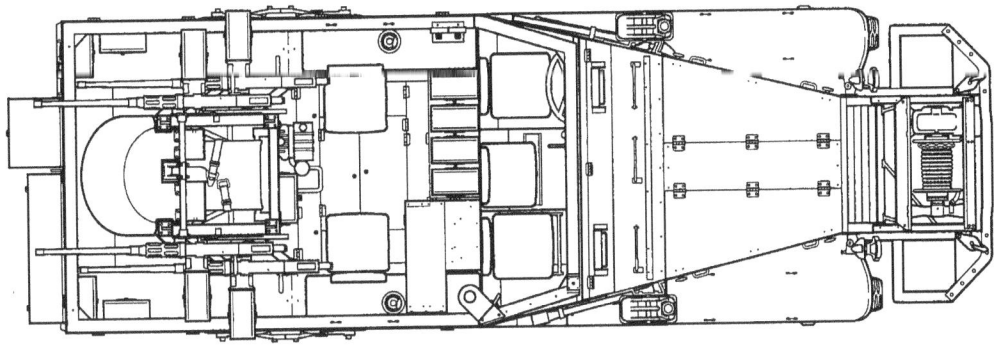

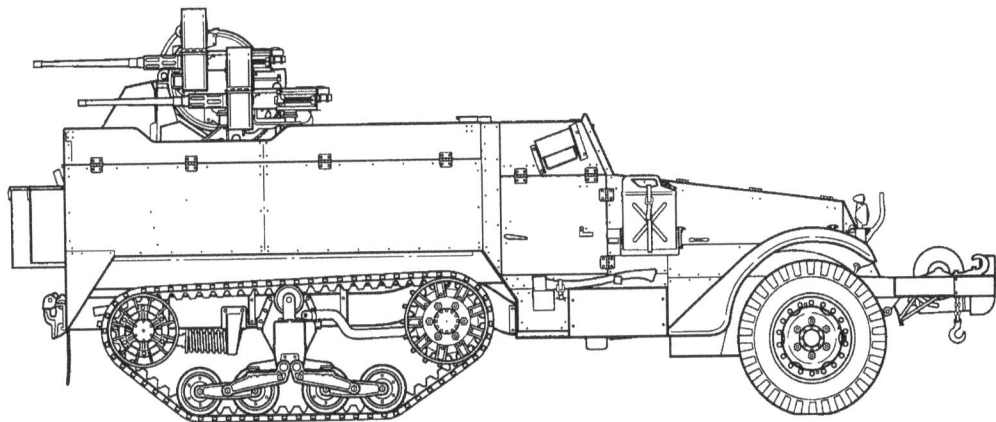

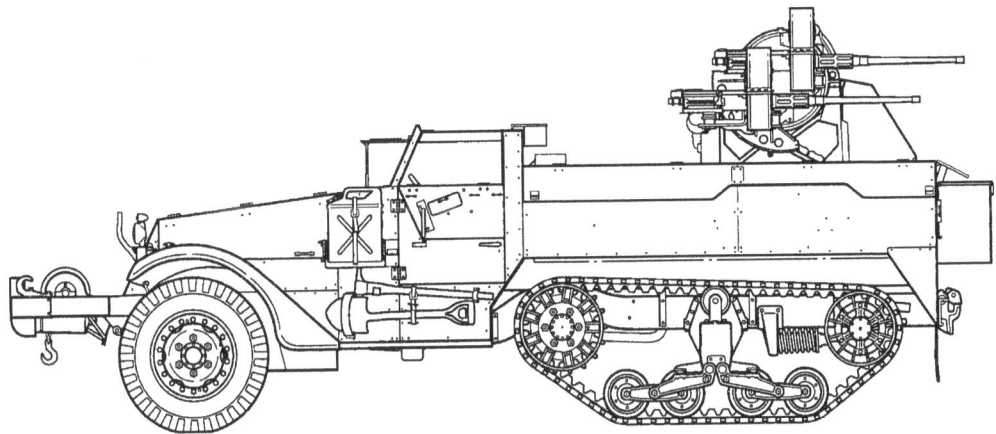

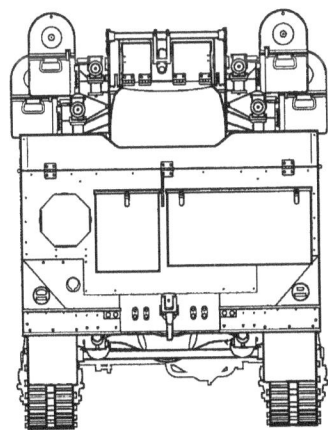

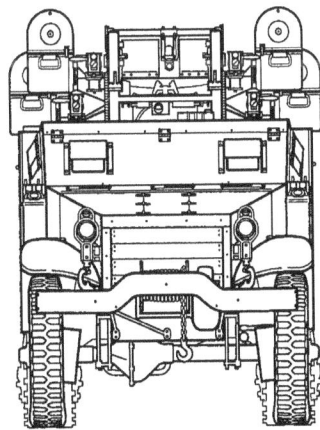

Scale 1:48

©M. Duplessis

Half-Track Multiple Gun Motor Carriage M16

These photographs from the Engineering Standards Vehicle Laboratory in Detroit, Michigan were dated 8 February 1944. They show the half-track multiple gun motor carriage M16, Ordnance serial number 2654, manufactured by the White Motor Company.

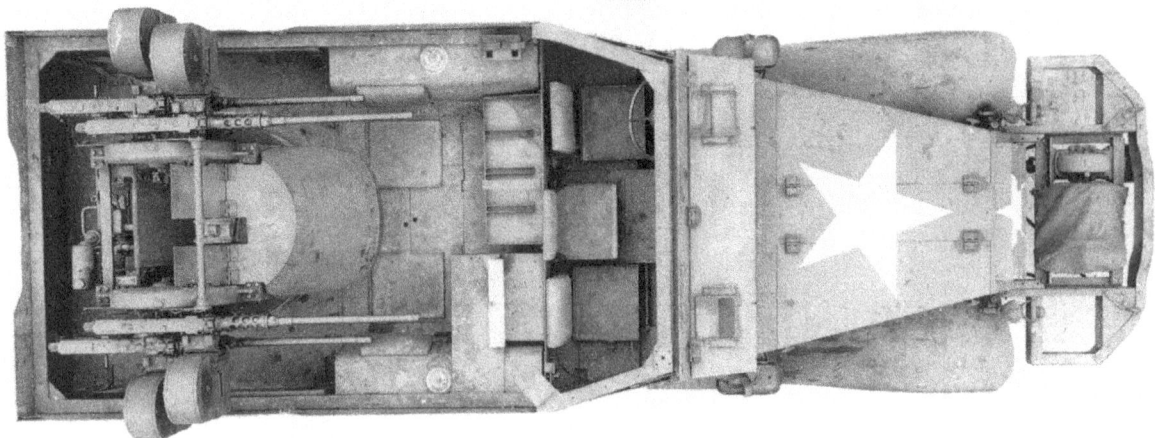

Above is a rear view of half-track multiple gun motor carriage M16, serial number 2654. The rear vehicle stowage and the rear of the gun mount are clearly visible. Also note the late model half-track taillights. Below at the left is a top view of the fully stowed half-track multiple gun motor carriage M16. The need for the cutouts in the folding side and rear armor is obvious. Details of the multiple .50 caliber machine gun mount M45D can be seen in the illustration at the bottom right.

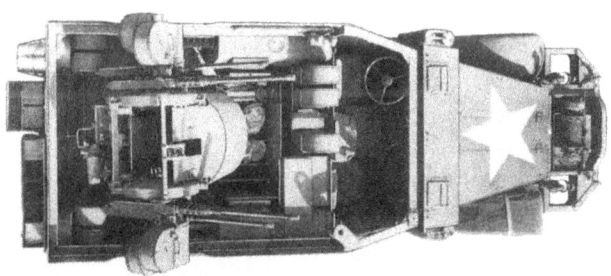

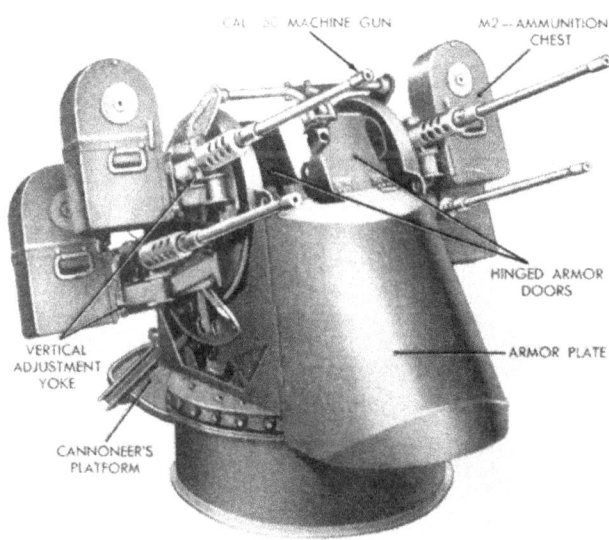

CAL .50 MACHINE GUN

M2—AMMUNITION CHEST

HINGED ARMOR DOORS

ARMOR PLATE

VERTICAL ADJUSTMENT YOKE

CANNONEER'S PLATFORM

The half-track multiple gun motor carriage M16 above is at Aberdeen Proving Ground during its evaluation. Below is a photograph of the half-track multiple gun motor carriage M17 manufactured by the International Harvester Company. Note the smooth armor and distinctive fenders of the half-track personnel carrier M5 on which it is based.

A total of 1000 M17 multiple gun motor carriages based upon the half-track personnel carrier M5 were built at the International Harvester Company. This vehicle utilized the same M45D mount with the four .50 caliber machine guns as the M16. Production of the M17 ran from December 1943 through March 1944. All of these vehicles were allocated to the Soviet Union under the Lend-Lease program.

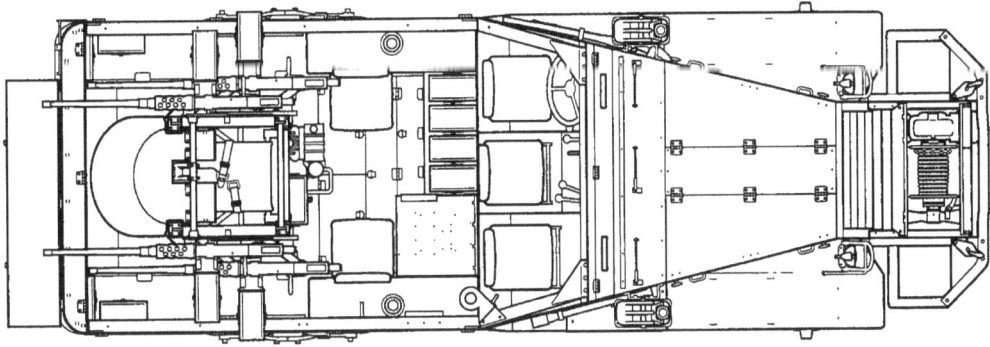

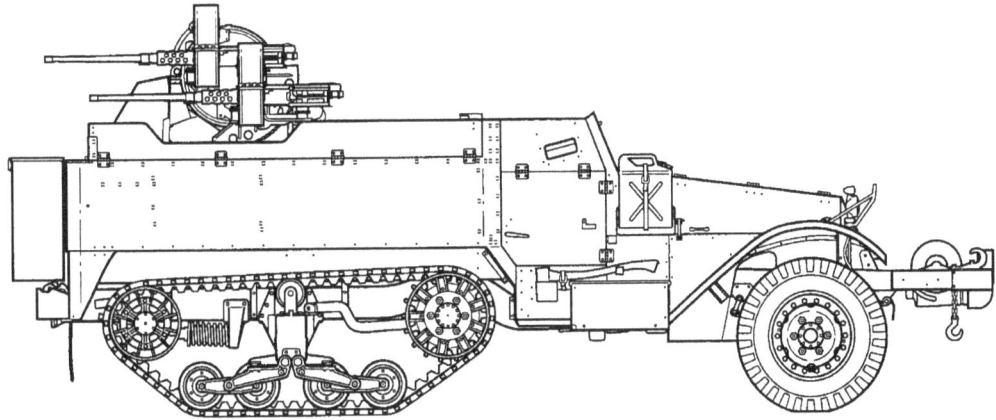

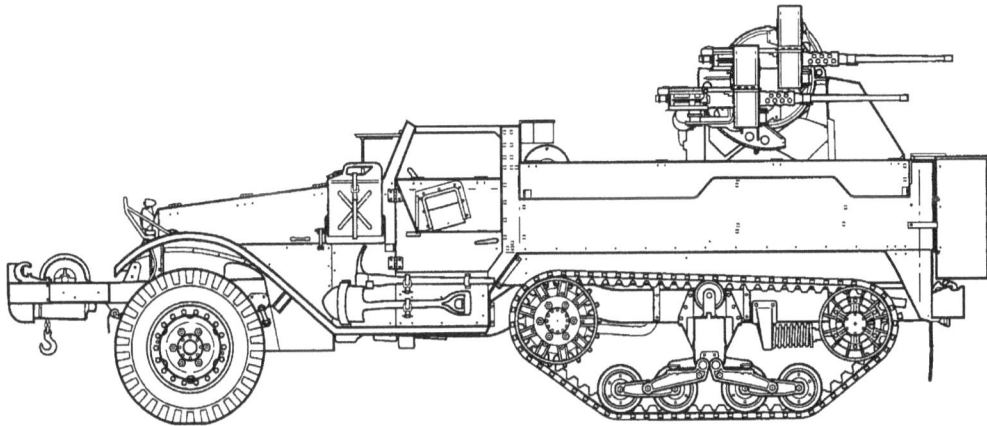

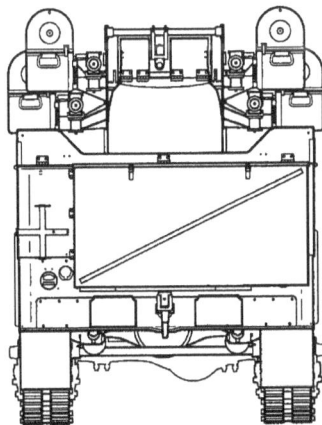

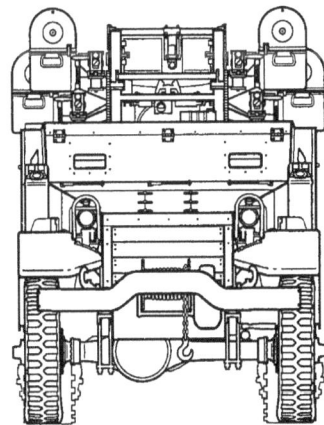

Scale 1:48

Half-Track Multiple Gun Motor Carriage M17

The half-track multiple gun motor carriage T10 appears in the photographs above and below. Above, the twin 20mm Oerlikon guns are at maximum elevation. The 20mm ammunition magazines are in place in the lower view. Half-track car M2, registration number 4011370, was a real workhorse at Aberdeen providing the chassis for a number of pilot vehicles.

When authority was granted for the development of the multiple gun motor carriage T1, either the .50 caliber machine gun or the 20mm gun was authorized to arm the vehicle. The immediate availability of the machine gun resulted in its selection. The success of the T1 renewed interest in the development of a motor carriage armed with the 20mm weapon. On 24 July 1941, OCM 17022 recommended the development of the multiple gun motor carriage T10. The weapons under consideration were the Oerlikon 20mm gun Mark IV and the Hispano-Suiza 20mm automatic gun AN-M1 or AN-M2. After consideration, the Oerlikon Mark IV gun was selected for the new self-propelled weapon. Like the multiple gun motor carriage T1, a power operated, aircraft type, turret was specified with elevation and traverse rates of at least 40 degrees per second. Gun mounts were designed by the United Shoe Machinery Corporation, R. Hoe and Company, and the W. L. Maxson Company. The designs were designated as the twin 20mm gun mounts T4, T8, and T17 respectively. Once again, the Maxson design proved to be superior. Modified from the earlier Maxson mounts it was lighter in weight and could be fitted with either two 20mm guns or four .50 caliber machine guns.

The pilot T17 mount was delivered to Aberdeen Proving Ground in October 1942 where it was installed on our old friend half-track car M2, registration number USA 4011370. This was the same vehicle that had served as the multiple gun motor carriage T1E2 and as the carrier for the four gun .50 caliber T61 machine gun mount during its initial tests. The combination was now dubbed the multiple gun motor carriage T10.

The half-track twin 20mm gun motor carriage T10E1, based upon the half-track personnel carrier M3, appears on this page. Note the folding armor on the sides and rear to increase the maximum depression of the 20mm guns.

In May 1943, the Antiaircraft Command recommended that 110 twin 20mm gun mounts T17 be procured after faults noted in the tests of the pilot were corrected. These mounts were designated as the twin 20mm gun mounts T17E1 and were to be mounted on modified multiple gun motor carriage M16 chassis and were designated as the twin 20mm gun motor carriage T10E1. Four pilot mounts were built and one was installed on the pilot M16 chassis at Aberdeen Proving Ground. Another mount was furnished to the British under the Lend-Lease program. The tests at Aberdeen and at the Antiaircraft Artillery Board were satisfactory, but the program was halted until suitable self-destroying ammunition could be developed for the 20mm gun. In the meantime, production of the T10E1 had started at the White Motor Company. The 110 vehicles requested were produced during March 1944. As mentioned previously, 109 of these vehicles were converted to M16 multiple gun motor carriages.

The twin 20mm gun mount T17E1 can be seen at the right installed in the half-track. This photograph at Aberdeen Proving Ground was dated 11 January 1944. At the bottom right is a view of the Oerlikon 20mm Mark IV gun.

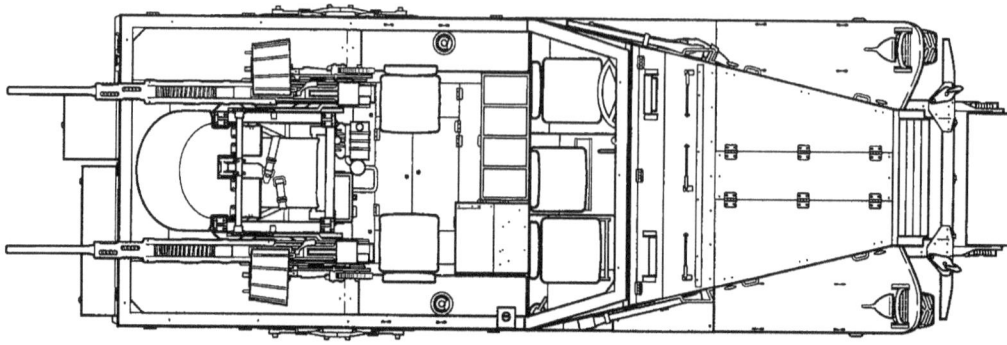

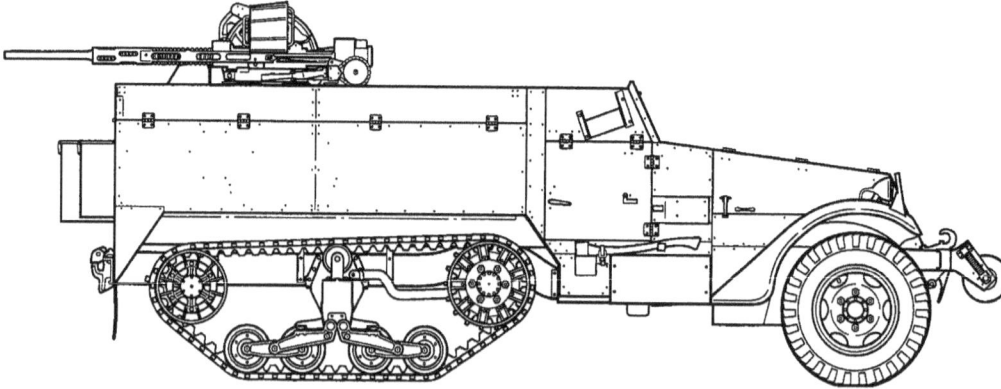

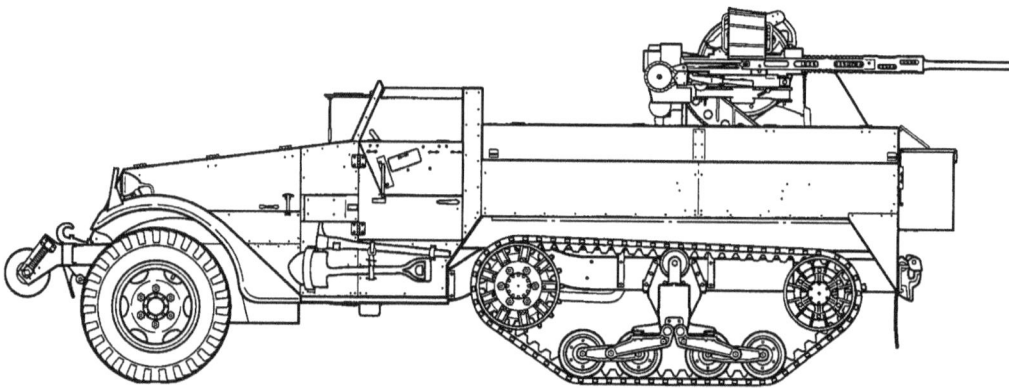

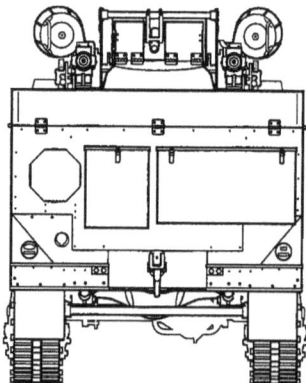

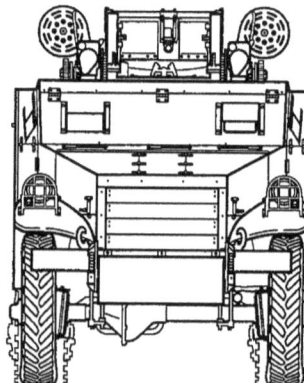

Scale 1:48

Half-Track Twin 20mm Gun Motor Carriage T10E1

Above is a photograph of the half-track twin 20mm gun motor carriage T10E1 at Aberdeen Proving Ground on 3 January 1944. The ammunition magazines are mounted on the 20mm guns. Two views of the production half-track twin 20mm gun motor carriage T10E1 appear at the right and below. Note the roller on the earlier T10E1 has been replaced by a front mounted winch and mine racks are installed on the sides of the vehicle. It also has late production features such as the double coil spring loaded idler and demountable headlights.

The Elco B-6 turret is shown at the left above and below mounted on a PT boat. At the right are photographs of the modified mount installed on half-track car M2, registration number 4012073, with an extended body. Note that the .50 caliber machine guns on the PT boat mount were M2 aircraft weapons. On the half-track installation they were replaced by .50 caliber M2HB machine guns.

An unusual multiple gun motor carriage was produced by the installation of a Navy experimental gun mount on a modified half-track personnel carrier M2. This was the vehicle that previously served as one of the pilots of the multiple gun motor carriage T28 with Registration Number 4012073. The body of the half-track was extended to provide space for the new armament. The mount was the Elco turret B-6 developed for use on PT boats. It was armed with four 20mm Oerlikon guns and two .50 caliber machine gun. Although it would have produced tremendous firepower, the vehicle was badly overloaded and the project was terminated without further development.

A front view of the half-track with the Elco mount is at the right.

The half-track 40mm gun motor carriage T54 appears in these photographs. At the top and bottom, the vehicle is being evaluated at Aberdeen Proving Ground on 14 July 1942. Below, the 40mm gun is at its maximum elevation.

At a conference on 25 June 1942, the decision was taken to investigate the feasibility of mounting the 40mm gun M1 on a half-track chassis for use as a self-propelled antiaircraft weapon. Two pilots were authorized for construction and OCM 18508, dated 16 July 1942, assigned the designation 40mm gun motor carriage T54. In order to expedite the program, the two pilots were assembled on the chassis of the half-track personnel carrier M3 without the side and rear armor. The pilots were completed by the Firestone Tire and Rubber Company in about two weeks. The gun mount consisted of the top carriage and the mating base from the 40mm gun carriage M2. The new mount was designated as the 40mm gun mount T5. The only fire control equipment installed were the standard direct fire sights. Tested at Aberdeen Proving Ground, the T54 was not as stable as the standard antiaircraft gun carriage, but it was considered to be worth further development after modification. The Antiaircraft Artillery Board also tested the new vehicle in comparison with the multiple gun motor carriage T28. Although they did not approve either weapon, they did make suggestions for additional modifications.

In August 1942, it was recommended that the T54 be modified to lower the gun mount as far as possible to improve stability and to install outriggers that could be quickly emplaced to further stabilize the vehicle. Armor protection also was required and the installation of an M5 type lead setting sight was recommended. The original T54 gun motor carriage carried the Registration

153

The half-track 40mm gun motor carriage T54E1 is shown above and below. The photograph above was taken at Aberdeen Proving Ground on 12 October 1942. The vehicle is similar to the T54, but it was protected by armor shields. Jacks and outriggers were used in an attempt to stabilize the vehicle when firing.

Number 4028245. The modifications were installed on a half-track with the Registration Number 402852 and it was designated as the 40mm gun motor carriage T54E1. As completed, the T54E1 was fitted with a circular armor shield $3/16$ inches thick and jacks were installed at the corners of the vehicle to lift it off the springs during firing. Stowage arrangements were provided for equipment and 120 rounds of 40mm ammunition.

OCM 18698, dated 27 August 1942, which outlined the modifications required for the T54E1 also recommended the manufacture of additional pilots. The first was designated as the 40mm gun motor carriage T59 with a remote control director carried in a separate vehicle designated as the half-track instrument carrier T18. The second was the multiple gun motor carriage T60. Both the T59 and the T60 were the same as the T54E1 except for the director fire control of the T59 and two .50 caliber M2 machine guns installed coaxially with the 40mm gun on the T60.

Tests of the new pilots revealed that despite the outriggers and jacks, they still did not have the stability required for accurate antiaircraft fire. A spring block-out system also was evaluated without any improvement. Another problem was the increased weight which was

The half-track 40mm gun motor carriage T54E1can be seen at the left with the gun elevated.

The half-track 40mm gun motor carriage T59 appears above without the circular armor shield (left) and with the shield (right). The half-track multiple gun motor carriage T60 is shown in the three photographs below. Note the use of the jacks and outriggers to stabilize the vehicles above and below at the left.

The half-track 40mm gun motor carriage T59E1 can be seen above and below with the reduced armor at Aberdeen Proving Ground on 17 June 1943. Details of the gun mount installation are visible in the view below.

now approaching 20,000 pounds. To alleviate the latter problem, the circular armor shield was removed during the test program. Later, low side and rear armor panels were installed on the T59 and T60. With the latter change, the vehicles were designated as the 40mm gun motor carriage T59E1 and the multiple gun motor carriage T60E1. By this time, development was underway of new full track 40mm gun motor carriages and there was no further interest in the half-track vehicles. The development the T54, T54E1, T59, T59E1, T60, and T60E1 was suspended by Ordnance Committee action on 12 October 1942.

The half-track multiple gun motor carriage T60E1 is at the right with the outriggers and jacks installed.

The half-track instrument carrier T18 above was intended for use with the half-track 40mm gun motor carriage T59 or T59E1. Based upon the half-track personnel carrier M3, the armor around the rear compartment was extended upward to provide additional protection for the crew. The T18 was at Aberdeen Proving Ground on 15 March 1943.

In February 1942, the American Ordnance Corporation submitted a design for a twin 40mm gun motor carriage based upon a half-track chassis. In April, the Ordnance Department accepted their proposal to construct a pilot without expense to the Government and loaned the two guns to be installed on the vehicle. Once again, the new motor carriage was based upon the half-track personnel carrier M3, this time with the twin 40mm gun mount T9. This mount carried the two 40mm guns one above the other resulting in a very high silhouette. The complete vehicle was designated as the twin 40mm gun motor carriage T68. The pilot T68 was shipped to Aberdeen Proving Ground arriving in December 1942. The tests indicated that a major redesign would be required before satisfactory operation could be achieved. Since superior full track 40mm gun motor carriages were already under development, the Ordnance Committee recommended no further participation in the development or test of the T68. The project was terminated on 24 June 1943.

Below and above at the right are photographs of the half-track twin 40mm gun motor carriage T68 at Aberdeen Proving Ground on 19 December 1942. Screw jacks can be seen installed on the bogie in the left side view at the bottom right.

Above are two views of the experimental mine flail installed on the front of a half-track personnel carrier M3.

SPECIALIZED HALF-TRACK APPLICATIONS

As mentioned earlier, the half-track was adapted for a wide variety of tasks in addition to its original intended role as an infantry personnel carrier, reconnaissance vehicle, or prime mover. Among these were the use of the M3 half-track personnel carrier as an armored ambulance. The M3 also provided transportation for the engineer squad with space for their special tools and demolition supplies. One experimental application that also might have been of interest to the engineers was the attachment of a flail type mine exploder to the front of the M3. With the light armor of the half-track, It probably would only have been safe to use with antipersonnel mines. However, it was only an experiment.

Below, Sergeant Charles Feider demonstrates compliance with the Geneva Convention restriction on medical personnel using firearms. He was with the Medical Detachment of the 6th Armored Field Artillery Group in Arkansas on 12 November 1942.

The stowage chart for the engineer squad in the half-track personnel carrier M3 is shown at the right.

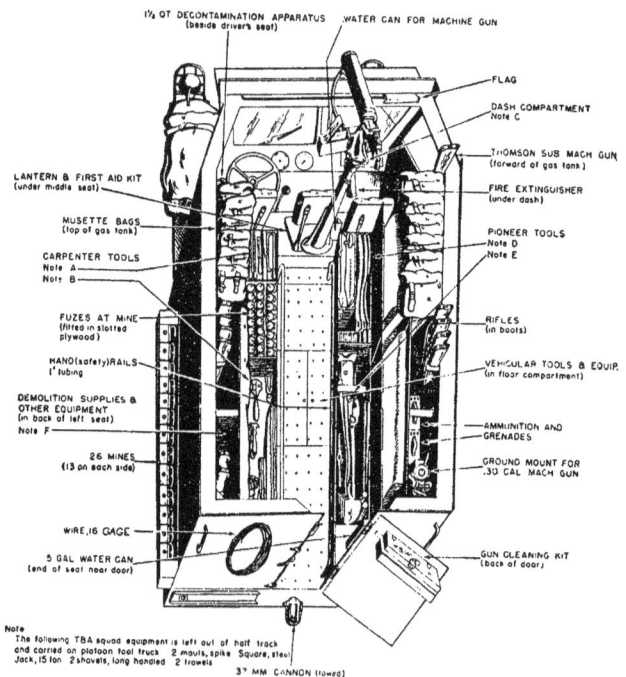

158

The half-track sonic deception carriers of the 3132nd Signal Service Company are shown above at Pine Camp, New York. Below at the right, the speaker is erected on one of the carriers. At the bottom of the page, the speakers are being erected on the carriers.

A Top Secret application of the M3A1 half-track personnel carrier was its use in the sonic deception program during the World War II. Two special companies were formed by the Signal Corps to perform this work. The first of these was the 3132nd Signal Service Company which was activated at Pine Camp, New York on 1 March 1944. The mission of the sonic deception units was to deceive the enemy by generating the tactical sounds of military operations up to and including those of an augmented armored division. The sound equipment, including the large speakers, was installed in the M3A1 half-track personnel carriers used by the 3132nd Signal Service Company. This unit served effectively in France and Germany.

The half-track radio carrier T17 is shown above based upon the half-track personnel carrier M3. The armored body has been removed and a truck cab and body installed on the half-track chassis.

The second sonic deception unit was the 3133rd Signal Service Company which was activated at Pine Camp on 21 June 1944. However, this unit carried the sound equipment in converted M10 tank destroyers. The 3133rd served in Italy during the last days of the war.

Another special application of the M3 personnel carrier was its modification to carry the SCR299 radio set. This set weighed about 5000 pounds. To reduce the total load on the vehicle, the armored body and hood were removed from the half-track and replaced with standard truck components. The 9 foot cargo body from the $2^{1}/_{2}$ ton truck was installed on the half-track chassis. Designated as the half-track radio carrier T17, four of these vehicles were shipped to the Desert Training Center at Camp Young, California for evaluation. The report from the Desert Training Center noted that the radio shelter HQ-17 was designed for a carrier with a 12 foot body resulting in a considerable overhang. They also noted that the T17 was inferior in performance to

the $2^{1}/_{2}$ ton GMC 6x6 truck when both were carrying a 5000 pound load of sandbags. Despite this, six modified vehicles were assembled at the Chester Depot. Designated as the half-track radio carrier T17E1, the floor was lowered on these vehicles and the radio equipment was rearranged. Provision also was made for installing a .50 caliber machine gun.

Another view of the half-track radio carrier T17 is at the right. Note the extreme overhang of the radio shelter at the rear of the vehicle.

The half-track T3 built by the Mack Manufacturing Corporation appears above at Aberdeen Proving Ground on 13 October 1941. Note the very long tracked rear suspension.

THREE-QUARTER TRACK VEHICLES

Efforts to increase the load carrying capacity and cross country mobility of the half-track vehicle resulted in designs with longer tracks reducing the space between the front wheels and the rear track suspension. Although they were officially designated as half-tracks, these vehicles were frequently referred to as three-quarter tracks. Some of these were similar in appearance to the German half-track vehicles of that period.

The Mack Manufacturing Corporation proposed a new half-track vehicle with a rear mounted engine. This design reduced the distance between the front wheels and the tracks to a minimum and was intended to carry 80 per cent of the vehicle weight on the tracks and only 20 per cent on the wheels. The front wheels were not driven, but the front wheel steering was coupled to differential steering of the tracks. The rear tracks and suspension were those of the M2 light tank. Designated as the half-track chassis T3, the vehicle was completed in October 1941 and delivered to Aberdeen Proving Ground. The test results showed excellent performance and good reliability, although a number of modifications were recommended.

The half-track 40mm gun motor carriage T1 appears at the right on 8 December 1941 at Aberdeen Proving Ground.

After the initial evaluation, the T3 was modified for the installation of a 40mm gun M1 and a Kerrison director. It was then designated as the 40mm gun motor carriage T1. The firing tests at Aberdeen showed interference between the gun and the director and concluded that it was impractical to install both on the same vehicle. The T1 was then returned to Mack and converted back to the basic T3 half-track chassis. By this time, other half-track projects were in progress and there was no further development of the T3.

The 40mm gun motor carriage T1 is shown in these three photographs. Note the interference between the gun barrel and the director at low elevations in the view below.

Above the half-track car T16 (left) manufactured by Autocar is compared with a standard half-track car M2 (right). Note the larger bogie wheels and the increase in ground contact length of the tracks.

In November 1941, drawings were prepared for a study of the installation of the 105mm howitzer M2A1 on the T3 half-track chassis. The proposed vehicle was to be designated as the 105mm howitzer motor carriage T34. However, the project was terminated when the T3 program was canceled in June 1942.

Another effort to obtain an improved half-track was based upon modification of the M2 half-track car. Designated as the half-track car T16, the frame was lengthened and a new track suspension was installed. Similar in design to the standard suspension, it was fitted with larger bogie wheels and the track width was increased from 12 to 14 inches. The ground contact length increased from 46.75 inches to 61.5 inches. This increased the ground contact area under the tracks reducing the ground pressure. Another feature of the T16 was a $1/4$ inch thick folding armor plate roof. The engine and power train remained the same as on the standard half-tracks. The increased weight of the longer frame and the armor top, slightly reduced the performance. As a result of the test program, it was concluded that the armor roof was not practical and the installation of the larger track and suspension was unsatisfactory. Further work on the T16 was stopped and the results of the tests were applied to the new half-track truck program.

The half-track car T16 below is at Aberdeen Proving Ground on 3 January 1942. The hinged section of the armor top is folded down to provide maximum protection around the .50 caliber M2HB machine gun.

In addition to the .50 caliber machine gun, two .30 caliber M1919A4 machine guns were mounted on the half-track car T16. Note the tripod mounts for ground use on the rear of the T16 above. Below, the front roof armor is folded back to improve visibility and to allow more freedom of movement for the .50 caliber machine gun.

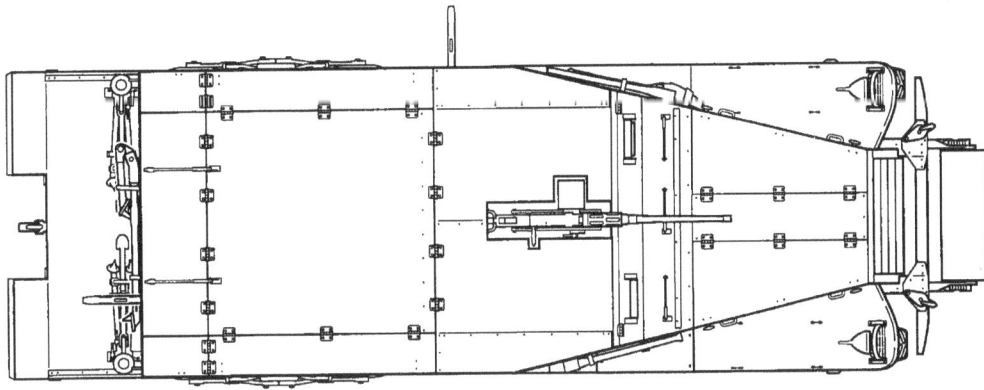

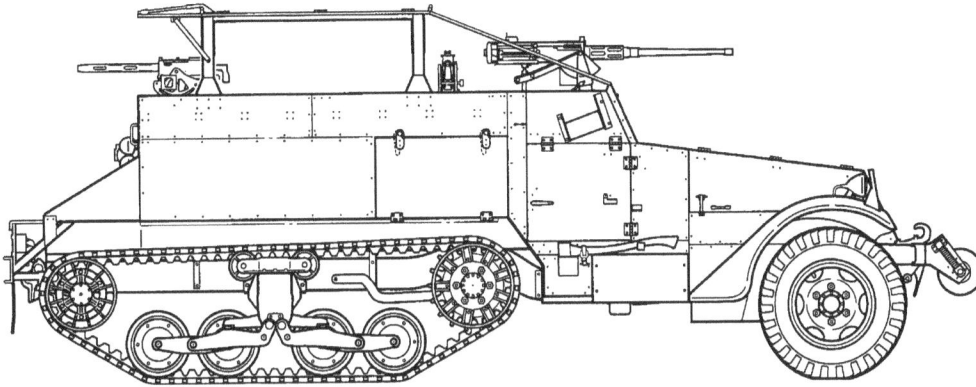

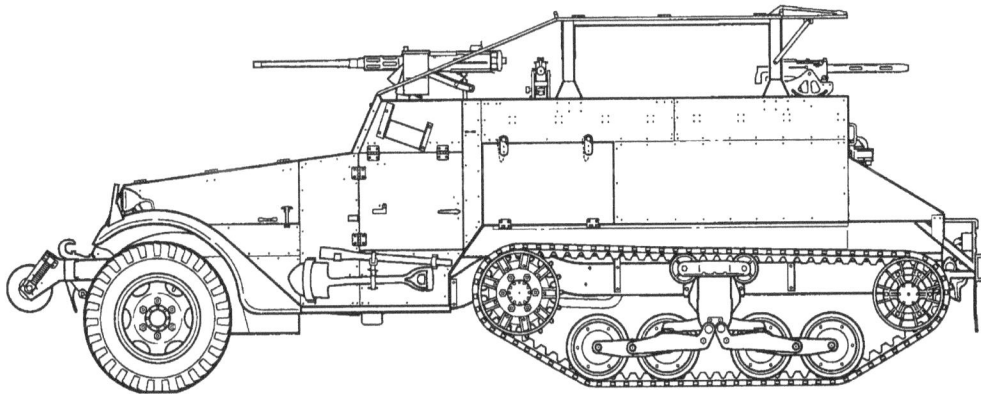

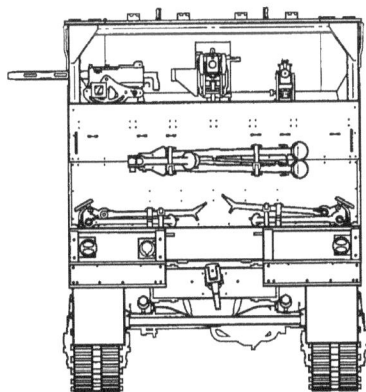

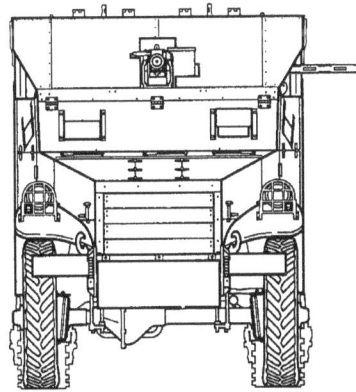

Scale 1:48

Half-Track Car T16

The roof armor on the half-track car T16 is closed above and open in the view below. Note that all of the machine guns are skate mounted on the gun rail encircling the inside of the vehicle. The T16 reflects its early construction with a fixed idler and the early headlights.

OCM17203, dated 11 September 1941, outlined the characteristics for the half-track truck T14. This proposed a vehicle weighing about ten tons including a payload of about 5600 pounds. It was to be armored on all sides and top for protection against .30 caliber ball ammunition and have a level road speed of 45 miles per hour and a constant speed of 25 miles per hour on a 60 per cent slope. All of this was required with a full payload plus a towed load such as a 105mm howitzer.

The Half-Track Vehicle Committee reviewed these requirements at a meeting on 20 October 1941 and concluded that the performance could not be achieved within the weights specified. As a result, the T14 project was terminated. At this same meeting, the Committee proposed the concurrent development of five new half-track designs. These were subsequently designated as the half-track trucks T15, T16, T17, T18, and T19 by OCM 17968 on 26 March 1942.

The military characteristics specified were similar for all five vehicles. They were intended to serve as prime movers for artillery pieces that weighed no more than 6500 pounds. They were expected to carry a payload of 6000 pounds including a 14 man crew and have all round armor protection against .30 caliber ball ammunition. Armament was to consist of a .50 caliber machine gun for use against aerial or ground targets. As initially proposed, the five vehicles differed mainly in the power train and suspension.

The half-track truck T15 was to use a White 24AX front mounted engine and a Spicer syncromesh transmission with a power shift. The front axle was to be powered. The tracks were driven from the front and were to be the same as on the half-track car T16.

The half-track truck T16 was to have a front mounted Hercules RXLD gasoline engine or it could be replaced by a General Motors 6-71 diesel. The transmission was to be a Spicer automatic torque converter.

The front axle was to be driven. The equalized rear suspension was to be fitted with 14 inch wide band tracks driven from the front.

The half-track truck T17 was to use a front mounted Hercules RXLD engine with a 4 speed Spicer syncromesh transmission. The front wheels were to be driven. The rear suspension and tracks were to be the same as on the half-track car T16 and were to be driven from the rear.

The half-track truck T18 was to be driven by a rear mounted General Motors 6-71 diesel engine with a 5 speed General Motors or Spicer syncromesh transmission fitted with a power shift. The front wheels were not to be driven. The proposed front driven tracks and suspension were from the light tank M3 with a trailing idler.

The half-track truck T19 was to be powered by a rear mounted White 24AX engine using a 5 speed General Motors or a Spicer syncromesh transmission with a power shift. The front wheels were not to be driven. A front drive equalized suspension with light tank M3 tracks was proposed without a trailing idler.

The T15 concept was dropped prior to the start of the development program because of its similarity to the T17. Since the T15 and the T17 had been proposed by the White Motor Company and the Autocar Company respectively, each received a contract to build one pilot of the T17. However, its features were combined with those of the T15. The T17 was now powered by the White 24AX engine. This was a 12 cylinder horizontally opposed, liquid-cooled, gasoline engine developing 210 horsepower at 2800 rpm. An equalized torsion bar suspension was installed with 14 inch wide band block tracks which were driven from the rear. Pilot number 1 was completed by White and delivered to the Armored Board at Fort Knox and the second was completed by Autocar and shipped to Aberdeen Proving Ground.

Below, half-track truck T17, pilot number 1, can be compared with the standard half-track personnel carrier M3 at Fort Knox.

Half-track truck T17, pilot number 1, appears on this page. Manufactured by the White Motor Company, it is at Fort Knox during its evaluation by the Armored Force Board. Interior details of the vehicle are visible in the upper photograph. Also, note the front mounted winch.

The second pilot half-track truck T17 appears here during tests at Aberdeen Proving Ground on 25 June 1943. Manufactured by Autocar, it can be compared with the White pilot on the previous page. Note that pilot number 2 has a provision for a ring mounted machine gun over the assistant driver's station.

The half-track truck T16 pilot is shown above and below. This vehicle, manufactured by the Diamond T Motor Car Company, was photographed at Aberdeen Proving Ground on 25 June 1943. The only armament installed is a single pedestal mounted .30 caliber M1919A4 machine gun.

Two pilots of the half-track truck T16 were manufactured by the Diamond T Motor Car Company and tests began at Aberdeen Proving Ground on 1 July 1943. The equalized volute spring suspension was fitted with 12 inch wide band type tracks which were driven from the front. As originally proposed, it was powered by the Hercules RXLD six cylinder, water-cooled, gasoline engine developing 174 horsepower at 2600 rpm.

Because of the policy restrictions on the use of diesel engines, the T18 concept was canceled.

The half-track truck T19 can be seen above and below at the General Motors Proving Ground on 18 December 1942. Built by the Mack Manufacturing Corporation, the cooling louvers of the rear mounted engine are visible in the photograph below.

Mack Manufacturing Corporation completed two pilots of the half-track truck T19. These vehicles were tested at Aberdeen Proving Ground and at the General Motors Proving Ground. As built, the T19 was driven by the Continental R6572 six cylinder, water-cooled, gasoline engine that produced 215 horsepower at 3000 rpm. Unlike the other pilot vehicles, the engine was installed at the rear. A Mack 4 speed syncromesh transmission was used. The equalized volute spring suspension was fitted with $11^9/_{16}$ inch wide light tank tracks and driven from the front. This was the only one of the new half-track trucks that did not have powered front wheels.

The report of the comparative tests of the half-track trucks concluded that they were unsatisfactory and recommended the termination of the program. The main reason was that the user was no longer interested in half-track vehicles. The artillery was now using high speed tractors as prime movers and other users also were shifting to full track vehicles.

At the right, the half-track truck T19 is at Aberdeen Proving Ground on 26 October 1943. The canvas top has been installed on the vehicle.

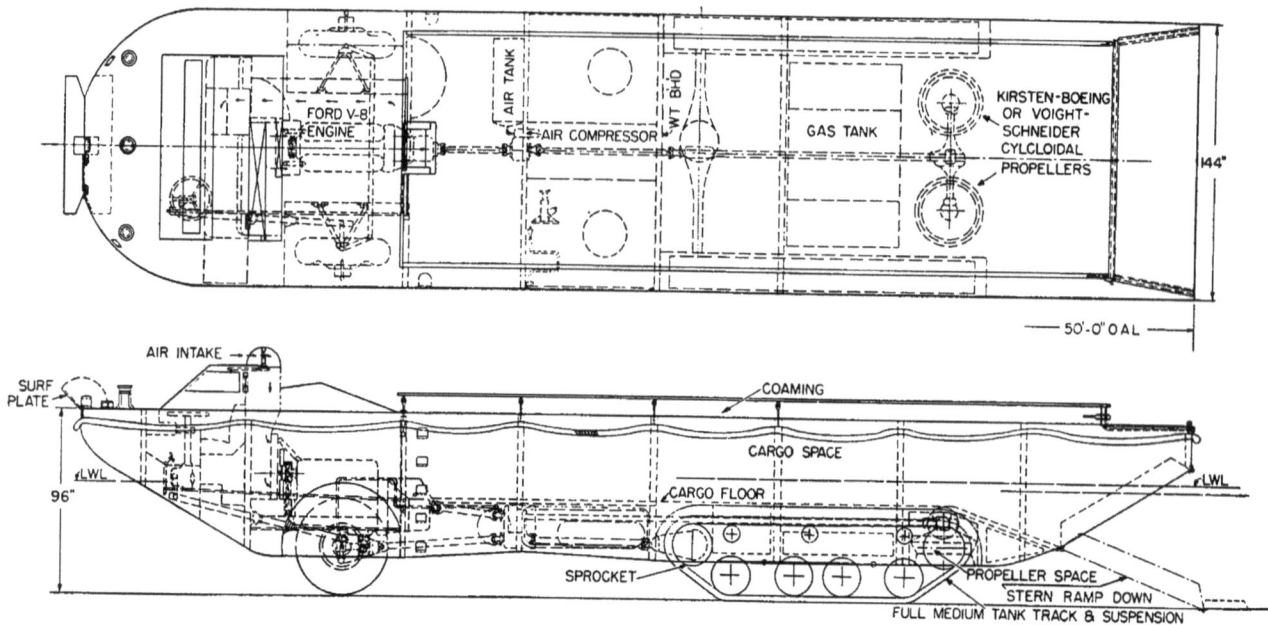

The drawing above shows the final version of the proposed Pelican amphibian with two front wheels and the full medium tank track and suspension.

HALF-TRACK AMPHIBIANS

By July 1942, the early work by the National Defense Research Committee (NDRC) on the DUKW amphibious $2\frac{1}{2}$ ton truck had indicated the desirability of a larger vehicle. Amphibious conversions were studied of several larger trucks and it was noted that for payloads greater than six tons, wheeled vehicles had serious disadvantages. Studies were then initiated on a series of half-track amphibians with payloads of 2 to 25 tons. Referred to as the Pelican project, some of these concepts used tank tracks in combination with two or four front wheels. The final proposal from this program envisioned a vehicle 50 feet long powered by a 400 horsepower engine that would have a water speed of 8 miles per hour. The vehicle would weigh 20,000 pounds empty and 40,000 pounds with a full load. A stern ramp provided easy access and the cargo space was large enough for a 6 ton 6x6 truck. No pilots were built of these vehicles, but numerous scale models were used in the water tests.

In May 1944, the Ordnance Department requested the NDRC to undertake the design of a large three-quarter track carrier amphibian for ship to shore operations. The actual design work was carried out at Sparkman & Stephens, Inc. under contract with the Office of Scientific Research and Development (OSRD) in cooperation with the Ordnance Department.

The new half-track vehicle, as designed, had an estimated empty weight of 37,000 pounds and a payload of 30,000 pounds. It had a cargo floor space of 200 square feet and was fitted with a rear opening ramp. The power train consisted of the Continental R975-C4 engine with the General Motors Torqmatic transmission. This was the same combination used on the 76mm gun motor carriage M18. The independent torsion bar suspension carried the vehicle on five dual road wheels per side in the original design. Later this was increased to six. The 21 inch wide tracks had been developed for use on the 105mm howitzer motor carriage T87. The drive sprocket was at the front of the tracks and an idler was at the rear. The front wheels with 14.00 x 24 tires were power driven. Water propulsion was by two 28 inch diameter propellers. The estimated maximum speed was 30 miles per hour on land and 8 miles per hour in water. As mentioned earlier, the design was modified to add an additional road wheel per side. Another change was to the rear hull. This was modified to provide enough clearance for the vehicle to enter a landing ship tank (LST) from the water.

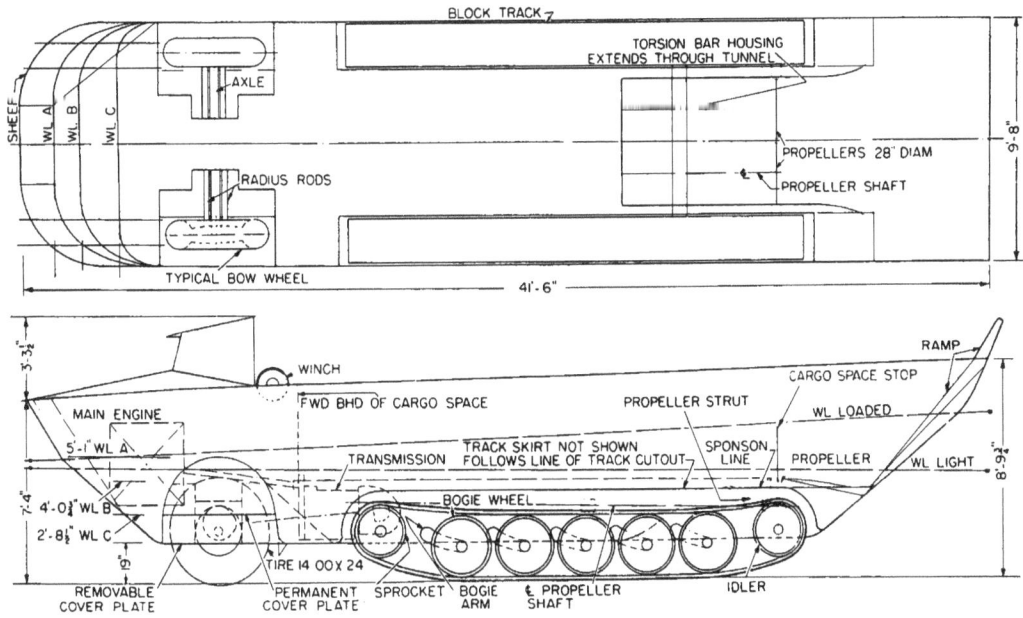

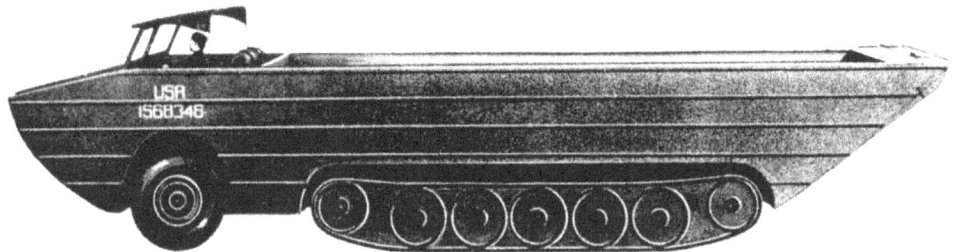

The original version of the 15 ton amphibian cargo carrier is shown in the drawing at the top of the page and in the photographs of the model above and at the right.

OCM 24252, dated 29 June 1944, recommended that three vehicles be procured and designated them as the half-track amphibian cargo carrier T32. Unfortunately, the Army Service Forces did not approve and the T32 program was canceled.

The revised configuration of the 15 ton amphibian cargo carrier T32 appears below. Note the six road wheels and the redesigned rear ramp to provide clearance for the vehicle to enter an LST from the water.

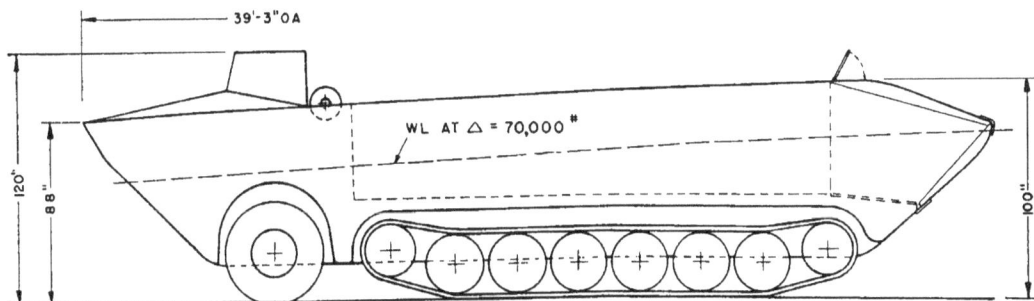

The semi-track configuration also was applied to a number of lightweight vehicles. Among these were some of the snow tractors. The snow tractor M7 was one such vehicle. Developed from the snow tractor T26 produced by the Tucker Sno-Cat Company in late 1942, a total of six were built. After a poor showing during their evaluation, most of these were modified to new test configurations such as the T26E1 and T26E2. A later version, designated as the T26E3, used a new track frame design. The latter was lengthened on the final version which became the T26E4 built by the Allis-Chalmers Manufacturing Company. The T26E4 was standardized as the snow tractor M7 in August 1943. They were reclassified as Limited Standard in November 1944.

The M7 was powered by the Willys MB Jeep engine. The band type tracks were 18 inches wide with a ground contact length of 60 inches. The front wheels were fitted with 4.00 x 15 tires. The front wheels could be replaced by skis for operation in deep snow. The skis were 9 inches wide and 70 inches long. When the front wheels were installed, the skis were mounted on each side of the hood to serve as mud guards. With a two man crew, the M7 had a gross weight of 3049 pounds. The M7 could tow the one ton snow trailer M19 which also could be fitted with either wheels or skis.

Another series of light half-track snow tractors began with the T27 built by the Seaman Motor Company. After the first vehicle, the construction was taken over by the Allis-Chalmers Manufacturing Company. Development continued through the T27E1, T27E2, T27E3, T27E4, and T27E5. None of them were ever standardized.

The snow tractor T29 was based upon the 6x6 $^3/_4$ ton vehicle with tracks around a rear driving wheel and a bogie. An non-powered front axle could be fitted with either wheels or skis. An improved version of this vehicle, based upon the $^1/_4$ ton 4x4 Jeep, was built by the International Harvester Company (IHC) and designated as the snow tractor T29E1. This vehicle was fitted with a more efficient track and a larger cargo body. The front wheels remained on the hubs even when the skis were installed. As on the T29, the powered front axle was replaced by a lighter weight non-powered version. With a crew of three, the T29E1 weighed 3840 pounds.

The T29E1 provided the basis for the development of the half-track litter carrier T28. On the T28, the non-powered front axle with 6.00 x 16 tires was replaced by a powered axle carrying wheels with 7.50 x 16 tires. The transfer case also was modified to compensate for the 2 to 1 reduction in the rear axle unit and for the differences in the rolling radii of the tires in the front and rear. A more rugged track was developed to improve the operating life. The cargo body was redesigned to carry two litters on the right side extending into the driving compartment. Thus the assistant driver was eliminated when carrying litters. A seat on the left side was provided for three sitting cases and the attendant was seated on the step that unfolded out of the rear door. Pilot models of the half-track litter carrier T28 were shipped to Aberdeen Proving Ground and Fort Riley in December 1944. The tests revealed some problems that could probably be corrected. However, the Army Ground Forces made the decision to terminate the program and the project was closed.

The snow tractor T29E1 fitted with skis can be seen in the photograph above. Below is its descendant, the half-track litter carrier T28. Both show their strong relationship to the Jeep.

PART III

THE HALF-TRACK GOES TO WAR

The half-tracks on this page were operating during the training period prior to Pearl Harbor. At the top right, this half-track car M2 is on maneuvers near Leesville, Louisiana during September 1941. Note the three .30 caliber M1917A1 machine guns on this early vehicle.

HALF-TRACK OPERATIONS DURING WORLD WAR II

Among the combat vehicles in the U. S. Army during World War II, none had longer service than the half-track. In fact, some versions continued to serve until long after the Korean War. Their early production prior to the attack on Pearl Harbor was essential for the training of the armored units in the rapidly expanding army.

The half-track cars and personnel carriers provided valuable service in the first days of the war during the defense of the Philippines against the Japanese. The 50 T12 75mm gun motor carriages that had been rushed to the islands prior to the attack on Pearl Harbor were the only self-propelled guns available to the forces commanded by General MacArthur. Fortyeight of these guns were used to arm three provisional self-propelled artillery battalions. Each battalion consisted of four 4 gun batteries. These units were invaluable during the fighting on the Bataan peninsula. Some of the T12s were captured after the surrender and were taken into service by the Japanese.

The first American offensive operation of the war was the invasion of the island of Guadalcanal in August 1942. Once again, the 75mm gun motor carriage M3 was in action, this time with the U. S. Marine Corps. The Special Weapons Battalion in the Marine Division was equipped with 12 M3 75mm gun motor carriages. Used as self-propelled artillery, the M3 was employed effectively throughout the fighting on Guadalcanal. It served with the Corps during this and following campaigns until the latter part of the war.

Below, the half-track at the left is being used as the prime mover for a 75mm gun. The M2 half-track car at the right is serving as a platform to review the troops. Note the World War I type helmets in all of these photographs.

Above, half-tracks are fording a shallow stream during early training operations.

Below, an early production M2 half-track car from C Company, 194th Tank Battalion is operating on the island of Luzon during the defense of the Philippines against the Japanese.

At the right, this half-track 75mm gun motor carriage T12 was captured by the Japanese and taken into service. When American troops returned to the Philippines, it was knocked out and shoved off of a bridge into the water. Note the early gun shield characteristic of the T12.

The photographs above and below show the half-tracks in operation during maneuvers in Louisiana during August 1942.
At the left is an early half-track car M2 during training. The many shortages of equipment at that time is reflected in the dummy machine gun mounted on the gun rail.

General George S. Patton Jr. always had a great affection for customized vehicles. During his command of the First Armored Corp at the Desert Training Center near Indio, California in 1942, he had this half-track personnel carrier modified for use as a command vehicle. At that time, he was a major general, hence the two stars on his flag.

Above are the half-track 75mm gun motor carriages M3 of a tank destroyer company at Camp Hood, Texas during 1942. Below are two additional views of the half-track 75mm gun M3 tank destroyers during their early training. Note these soldiers are still wearing the old World War I helmets.

Above, Battery A, 59th Armored Field Artillery Battalion stands inspection with their T19 half-track 105mm howitzer motor carriages at Camp Chaffee, Arkansas on 31 October 1942. Below, the left sketch shows the crew positions on the T19 ready for action. The two right sketches show the crew in the ammunition carrier and the T19 mounted and ready to move. The photograph reveals that the crew did not always follow this arrangement. The crew members are designated as follows: D. driver, C-S. chief of section, G. gunner, 1-6. Cannoneers

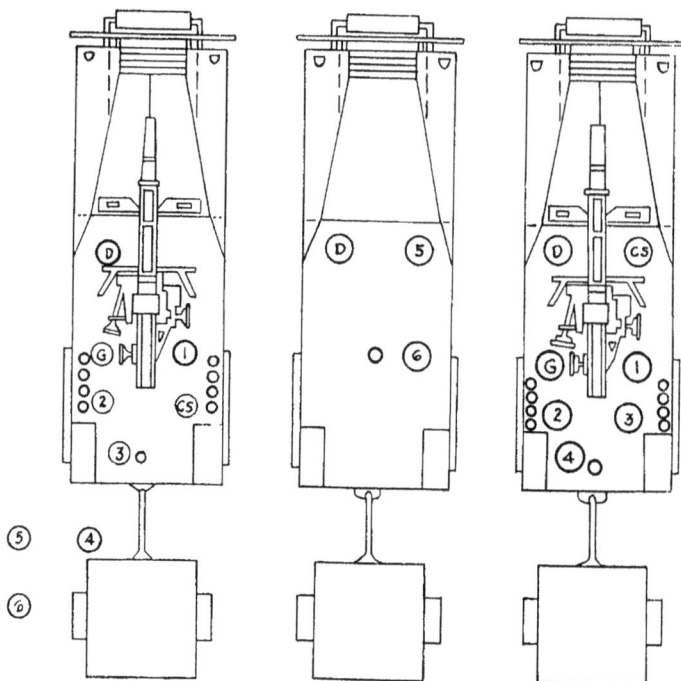

These photographs show the howitzer being loaded and fired in the half-track howitzer motor carriage T19. The unit is C Battery, 93rd Armored Field Artillery Battalion at Camp Chaffee, Arkansas on 31 October 1942. Note the pedestal mounted .30 caliber M1919A4 machine gun.

The photograph at the left shows the 105mm rounds stowed in the ammunition carrier.

At the left, a U.S. Marine Corps 75mm gun motor carriage M3 is being unloaded from a ship during the invasion of Guadalcanal in August 1942. Above, a half-track personnel carrier M3 is in Morocco after the invasion of North Africa. It is still fitted with the exhaust extension for deep water fording during the landing.

Above at the left, the 601st Tank Destroyer Battalion have their vehicles deployed near El Guettar, Tunisia on 23 March 1943. The M2 half-track car is in front and a 75mm gun motor carriage M3 is in firing position at the rear. Above at the right, another 75mm gun motor carriage M3 is operating in Tunisia. Note that this vehicle is fitted with mine racks. Below, an M3 half-track personnel carrier of the 1st Armored Division is towing its M3 37mm gun near Sidi bou Zid, Tunisia on 14 February 1943.

Above and below at the left are half-track multiple gun motor carriages T28E1 of the 443rd Antiaircraft Artillery Automatic Weapons Battalion in North Africa. Above at the right, a half-track multiple gun motor carriage M13 is camouflaged and ready for action.

In November 1942, the invasion of North Africa saw the participation of a wide variety of half-track vehicles. The half-track cars and personnel carriers as well as most of the self-propelled weapons on the half-track chassis were in action. The 443rd Antiaircraft Artillery, Automatic Weapons Battalion was part of the Western Task Force under Major General George S. Patton Jr. Some of their 78 T28E1 multiple gun motor carriages had even been stowed on the top deck of the invasion ships as antiaircraft protection. The T28E1 proved to be a very effective weapon against low flying aircraft such as the Stuka dive bomber. It served throughout the campaign in North Africa and the invasion of Sicily. On 12 September 1943, the Battalion was reorganized. The new Tables of Organization specified 32 T28E1 multiple gun motor carriages and 32 M13 multiple gun motor carriages. This, of course, greatly reduced the firepower available. Since the 443rd was the only unit armed with the T28E1, they requested that the extra vehicles be retained for use as replacements and spare parts. After battle losses and the transfer of eight vehicles to the French, 21 surplus T28E1s remained. With the new organization, the 443rd served in Italy and then participated in the invasion of southern France. At that time, they upgraded their M13s by replacing the twin .50 caliber M33 mounts with the four gun M45 mount giving them the same firepower as the new M16 multiple gun motor carriage. The 443rd continued to serve until the end of the war in Europe. At that time, they were in Germany and 26 of the original T28E1s were still operational.

At the left, the upper photograph shows the half-track multiple gun motor carriage T28E1 of A Battery, 443rd AAA AW Bn. ready for action in Italy. Note the six kill marks on the side of this vehicle. In the lower view at the left, a T28E1 of the 443rd is undergoing maintenance while protecting the St. Raphael air field in southern France on 17 August 1944.

Above, the engine of a half-track 75mm howitzer motor carriage T30 is being inspected shortly after landing near Algiers in North Africa. Perhaps it got a little too wet. The date is 8 November 1942. The T30s at the right are in action near Licata, Sicily. Below, another 75mm howitzer motor carriage T30 is firing on enemy positions at Anzio, Italy on 8 March 1944.

Above, a half-track multiple gun motor carriage M15 is ready for action in the Lungo area, Italy. At the left, this M15 is near Capra, Italy on 20 November 1943. The half-track combination gun motor carriage M15A1 appears in the two bottom photographs at Hoeville France (left) and Sedan, France (right). They were dated 3 November 1944 and 27 December 1944 respectively. Note the folding upper front armor on the M15A1.

Above, a camouflaged half-track 105mm howitzer motor carriage T19 is still in active service following the invasion of southern France in August 1944. At the right, the 37mm gun and mount from the 37mm gun motor carriage M6 has been installed in a half-track car M2. This vehicle of the 41st Armored Infantry, 2nd Armored Division was photographed in England prior to the invasion of Normandy. Note the new stowage bin on the back of the half-track and the elevated mount for the .50 caliber machine gun.

The 75mm gun motor carriage M3 was used as a tank destroyer in North Africa, but it was soon replaced in that role by the full track 3 inch gun motor carriage M10. However, it continued to serve with some units as self-propelled artillery. As in the case of the T28E1, some of the expedient weapons based upon the half-track chassis continued to serve in Sicily, Italy, and even into southern France and Germany. Some units transferred from the Mediterranean Theater of Operations to England for the invasion of Normandy made field modifications to their earlier vehicles. This was particularly true of the 2nd Armored Division. The wheeled 37mm gun motor carriage M6 was a failure as a tank destroyer in North Africa. However, in some cases, its 37mm gun and shield were removed and installed in the rear compartment of the half-track car M2 as an expedient weapon.

At the right is another example of the gun mount from the 37mm gun motor carriage M6 being installed on the half-track car M2.

188

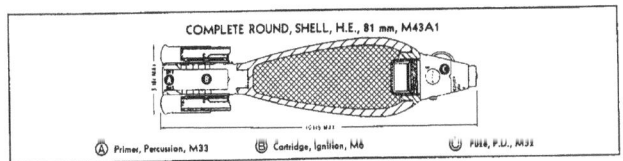

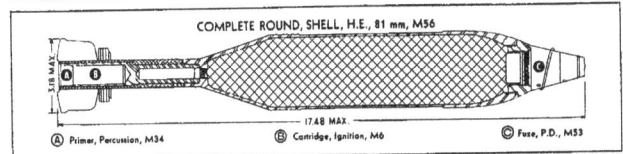

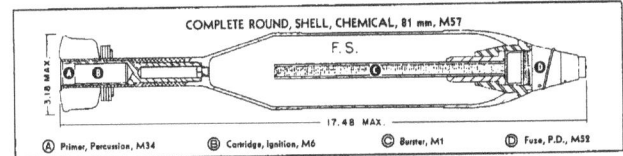

At the left are two photographs of the half-track 81mm mortar carrier M4 with the mortar mount modified to fire toward the front. These vehicles belonged to the 2nd Armored Division. The mortar is being fired in the lower view. Three types of 81mm mortar ammunition are shown in the drawing above.

Battle experience had shown that it would be preferable to have the mortar in the half-track mortar carrier M4 aimed toward the front instead of the rear. The 2nd Armored Division did not wait for the appearance of the half-track mortar carrier M21, but converted their M4s to fire toward the front.

The modified half-track mortar carrier is illustrated in the drawing below. Note that the vehicle has been upgraded to the late M4A1 standard with the spring loaded idler and the demountable headlights.

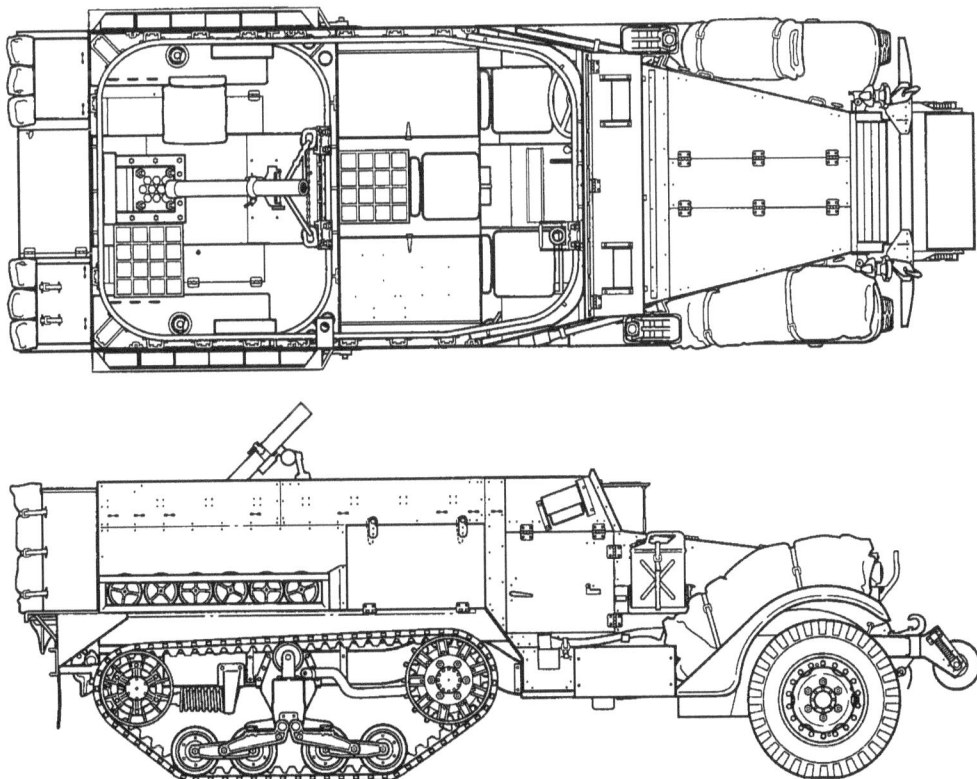

Scale 1:48

©M. Duplessis

Half-Track 81mm Mortar Carrier M4, modified by 2nd Armored Division

Above are two photographs of the "M16B" with the quadruple .50 caliber machine gun mount installed in the rear compartment of the half-track car M2. In the view at the right, the 377th AAA AW Bn. is in action in Normandy on 12 July 1944. A side view of the improvised vehicle is shown in the drawing at the bottom of the page.

One conversion reached the dimensions of mass production. Ordered by the Chief Ordnance Officer of the First U. S. Army, Colonel (later Major General) John B. Medaris, this was the installation of a large number of quadruple .50 caliber machine gun mounts on surplus half-tracks prior to D-Day in Normandy. According to the History of the U.S. Army in World War II there were 321 such conversions. However, the original First Army After Action Report indicates that 332 such vehicles were converted. These weapons were removed from their trailer mounts and installed in the rear compartments of the half-tracks. Unofficially dubbed the M16B, they were issued to twenty antiaircraft artillery battalions (16 each) previously armed with the towed weapon. They were employed in the beachhead phase of the invasion and when enemy aircraft activity decreased, they were used in the ground support role. These battalions supplemented the units equipped with the standard M16 multiple gun motor carriages for the remainder of the campaign in Europe.

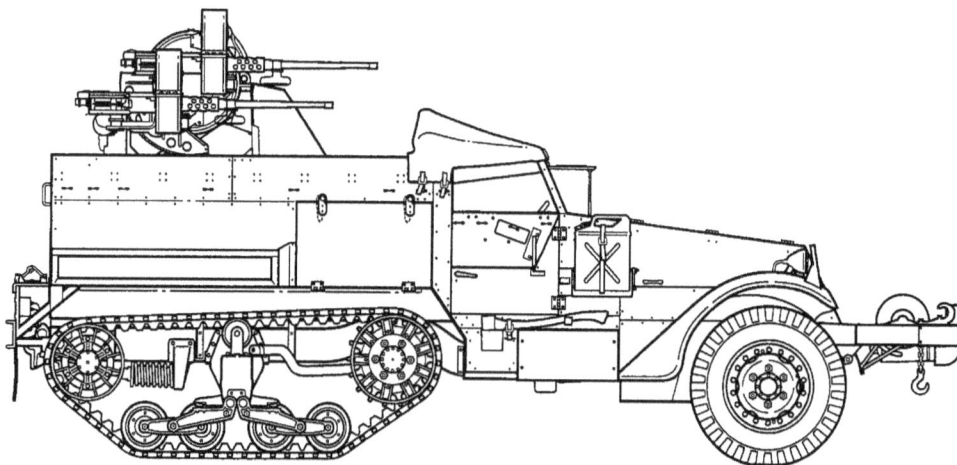

Scale 1:48

©M. Duplessis

Half-Track Multiple Gun Motor Carriage "M16B"

Above, a half-track multiple gun motor carriage M16 of the 447th AAA AW Bn. is ready for action near Neufchateau, Belgium on 1 January 1945. Note the five gallon fuel cans stowed on the front of the vehicle.

As mentioned before, the infantrymen preferred .50 caliber machine guns to the .30 caliber weapon specified on the M3 personnel carrier. The troops frequently made this change as described in "Hell on Wheels, the 2nd Armored Division" by Donald E. Houston.

"The armored infantrymen holding the line wanted to have all of the .30 caliber machine guns on their half-tracks replaced with the longer range .50 caliber guns. Some replacements had been made in England, but not all the guns had been changed. While holding the line, the men saw many American planes crash near their positions. The soldiers rushed to the crash, salvaged the .50 caliber machine guns and helped the pilots if they could. The regimental and division maintenance sections made mounts

so that the aircraft machine guns could be used in the infantry half-tracks. Procurement of supplies in this manner required a liberal interpretation of regulations about items found either on post or on the battlefield."

At the right, an improvised mount for two .50 caliber M2 aircraft machine guns has been installed on the front of a half-track.

191

Above, a half-track combination gun motor carriage M15A1 guards a bridge over the Rhine on 25 March 1945. Below, half-track personnel carriers M3 (left) and M3A1 (right) are advancing into Germany. At the bottom of the page, a half-track car M2A1 of the 61st Armored Infantry, 10th Armored Division follows its dismounted infantry into action in Germany on 17 April 1945.

Above, a U.S. Marine Corps half-track 75mm gun motor carriage M3 lands from an LST at Cape Gloucester on 26 December 1943. Another view of an M3 at Cape Gloucester appears at the right. The Marine Corps liked extra firepower on their half-tracks. Note the .50 caliber machine guns above and a .30 caliber M1919A4 machine gun is installed on a pintle mount on top of the shield at the right.

In the Pacific, the 75mm gun motor carriage M3 continued to support U. S. Marine Corps units during their island hopping campaign. The Army used a wide variety of half-tracks in the Pacific including the M16 and M15 series multiple gun motor carriages. A few of the latter were converted by replacing the 37mm gun and twin machine guns with a single 40mm gun M1. Referred to unofficially as the M15 Special, they were employed effectively in the Philippines.

The early Marine Corps half-track personnel carrier M3 below is in the Marshall Islands on 1 February 1944. At the right is a heavily armed Marine Corps 75mm gun motor carriage M3 on Saipan during June 1944.

These photographs show the "M15 Special" mounting the 40mm gun M1. The views above and at the right show the vehicle at Finschafen, New Guinea on 9 November 1944. Below, the "Special" is in action along the Villa Verde Trail, Luzon, in the Philippines on 8 May 1945.

All available half-track multiple gun motor carriages such as the M16 above were prepared for deployment to Korea.

KOREA

The outbreak of war in Korea during the Summer of 1950 resulted in a rush to provide combat vehicles to the troops deployed to Korea. Depots in Japan were searched for vehicles that could be refurbished for immediate use. These included various half-tracks. The M16 and M15A1 multiple and combination gun motor carriages were particularly valuable for their firepower in supporting infantry against the human wave attacks of the enemy. At that time, there was a shortage of 37mm ammunition, although there was an ample supply of 40mm rounds. As a result, 104 M15A1 combination gun motor carriages were converted in Japan by replacing the 37mm gun M1A2 and the twin .50 caliber machine guns with a single 40mm antiaircraft gun. Although it was originally referred to as the T19, OCM 33894, dated 15 September 1951, designated this vehicle as the 40mm gun motor carriage M34 and classified it as Limited Standard.

At the right is a half-track 40mm gun motor carriage M34 from B Battery, 140th AAA Bn. 40th Division in Japan on 30 September 1951.

Above, the standard half-track multiple gun motor carriage M16 on the right can be compared with an M16 fitted with the bat wings installed by the ordnance units. Note the greatly improved protection for the loaders.

The multiple gun motor carriage M16 also was effectively employed in the ground support role in Korea. However, only the gunner was protected on the gun mount. Crew members were particularly vulnerable when replacing the ammunition chests on the .50 caliber machine guns. To provide additional protection, ordnance units fabricated folding shields that could be extended on each side of the guns. Referred to as bat wings, they allowed the weapons to be reloaded without fully exposing the crew.

Later, when additional half-tracks were required with the quadruple .50 caliber mount, Bowen and McLaughlin, Inc. converted 1662 M3 half-track personnel carriers to multiple gun motor carriages. These were armed with the multiple .50 caliber machine gun mount M45F. This mount differed from the M45D by the addition of a six inch extension ring on the bottom to permit clearance of the guns over the non-folding sides of the M3 half-track personnel carriers. The mount also

was fitted with folding armor bat wings which differed somewhat from those fabricated in the field. A slip ring also was added to the mount to provide for an intercommunication system between the gunner on the mount, the squad leader in the vehicle cab, and to a point at the rear of the vehicle with connections for a field telephone. It also retained the rear door of the M3 personnel carrier. OCM 34189, dated 24 April 1952, designated the converted vehicle as the multiple gun motor carriage M16A1 and classified it as Substitute Standard.

Subsequently, the design features of the multiple gun motor carriage M16A1 were added to 419 M16s. These included the M45F mount with the bat wings, the slip ring under the mount, and the rear door. The latter required some rearrangement of the rear stowage boxes. OCM 34825, dated 11 May 1953, designated the converted vehicles as the multiple gun motor carriage M16A2 and classified it as Substitute Standard.

Below, the M16 on the left is from the 26th AAA AW Bn. of the 24th Division in Korea on 18 June 1951. The M16 at the right, from the 21st AAA AW Bn., is firing near White Horse Ridge, Korea in 1953. Both vehicles are fitted with bat wings.

The half-track multiple gun motor carriage M16A1 can be seen in the two views above. Since this vehicle was converted from the half-track personnel carrier M3, note the absence of the folding side and rear armor. The vehicle also retains the rear door of the M3. Details of the .50 caliber multiple machine gun mount M45F are visible below. Note the differences between the bat wings on the M45F and the field modified gun mount.

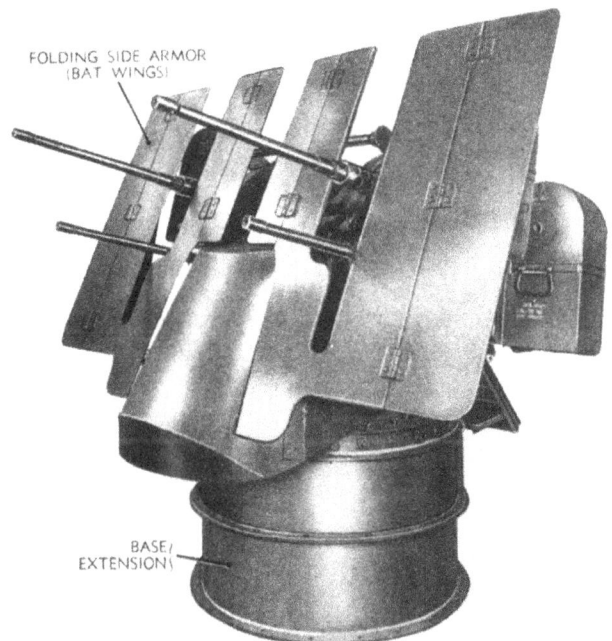

Two photographs of the standard half-track multiple gun motor carriage M16 in Korea appear below. At the left is an M16 from the 92nd AAA AW Bn. on 2 September 1950. At the right is an M16 from C Company 25th AAA AW Bn. on 30 March 1951.

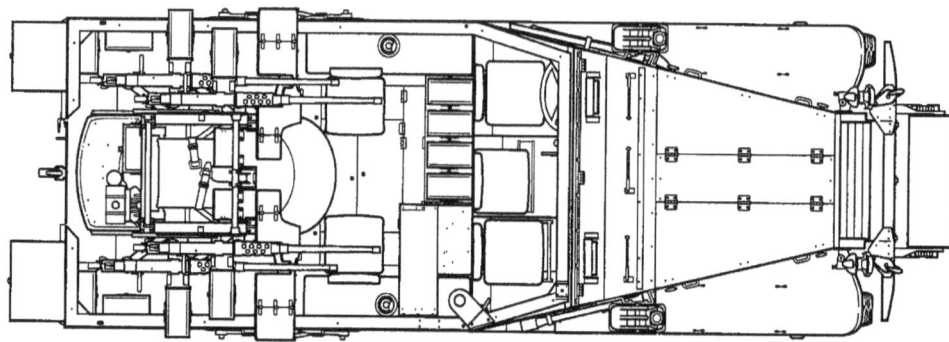

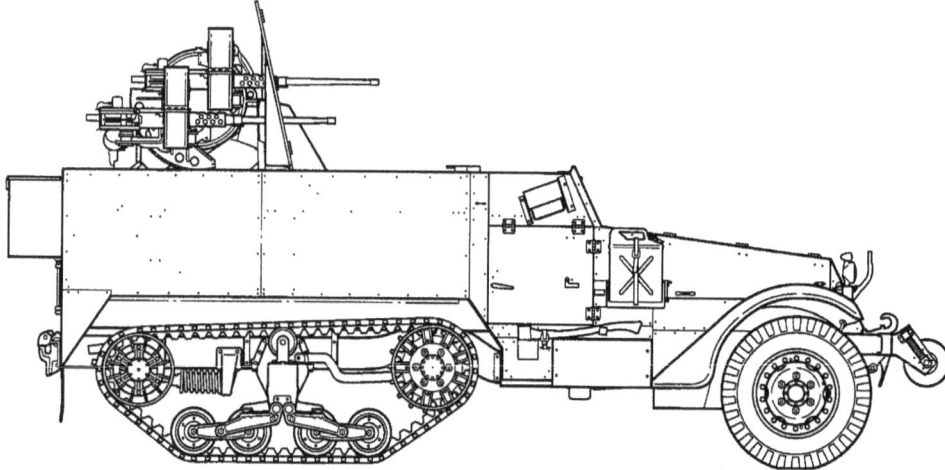

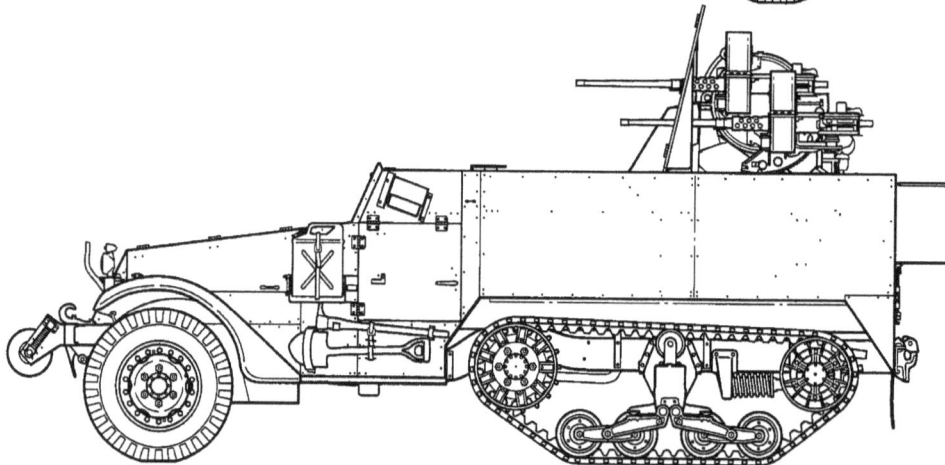

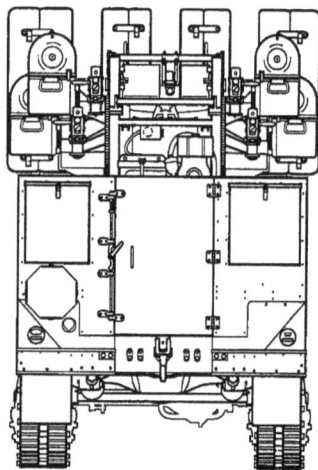

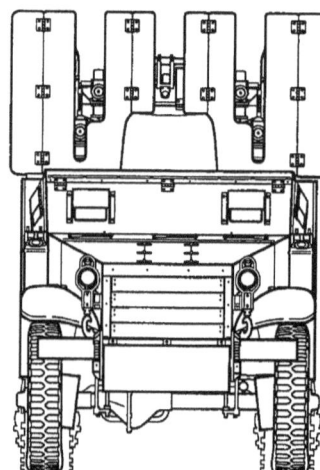

©M. Duplessis

Half-Track Multiple Gun Motor Carriage M16A1

After the 75mm gun motor carriage M3 was replaced in the tank destroyer battalions by later full track vehicles, many of the M3s were transferred to the British. Here, as shown above, they were employed as self-propelled artillery. Note the large ammunition supply in this vehicle. This M3 was in action in Italy during March 1945.

HALF-TRACKS IN FOREIGN SERVICE

During World War II, large numbers of all types of half-track vehicles were supplied to Great Britain and to the Soviet Union under the Lend-Lease program. Smaller numbers were transferred to other allied powers. As mentioned earlier, the half-tracks manufactured by the International Harvester Company were intended for foreign aid. These consisted of the half-track car M9A1, the half-track personnel carriers M5 and M5A1, and the multiple gun motor carriages M14 and M17. Most of the T48 57mm gun motor carriages were shipped to Britain or the Soviet Union. In addition, the Soviet Union received 100 of the M15 series combination gun motor carriages and all of the 1000 M17 multiple gun motor carriages. Additional half-track vehicles were transferred from the U. S. Army overseas to various allied nations.

At the right, a British half-track comes ashore at Les Andalouses, North Africa on 9 November 1942

After the end of World War II, the half-tracks were even more widely distributed. The were incorporated into the armed forces of many countries and widely modified by their new owners. In particular, Israel made wide use of the half-track as a personnel carrier and as the basis of several self-propelled weapons.

The half-track 57mm gun motor carriage T48 operating under its Soviet designation as the SU57 appears in the top photographs in Prague, Czechoslovakia in May 1945. Below, a Soviet half-track car M9A1 is towing a 6 pounder antitank gun in Prague.

At the right and below are three views of the half-track multiple gun motor carriage M17 in Soviet service. All of these photographs came from Ivan Bajtos via Steven Zaloga.

PART IV

REFERENCE DATA

The half-track 75mm gun motor carriage M3 with the early headlights and fixed rear idler was the configuration employed in the early days of the war at Guadalcanal and in North Africa.

Above is a half-track personnel carrier M3 with the early markings of the 2nd Armored Division. The half-track Car M2 below is on maneuvers at the Desert Training Center.

Above, a half-track 75mm howitzer motor carriage T30 is bogged down during operations in the Aleutian Islands. Below is a view of a half-track multiple gun motor carriage M13, also during training operations.

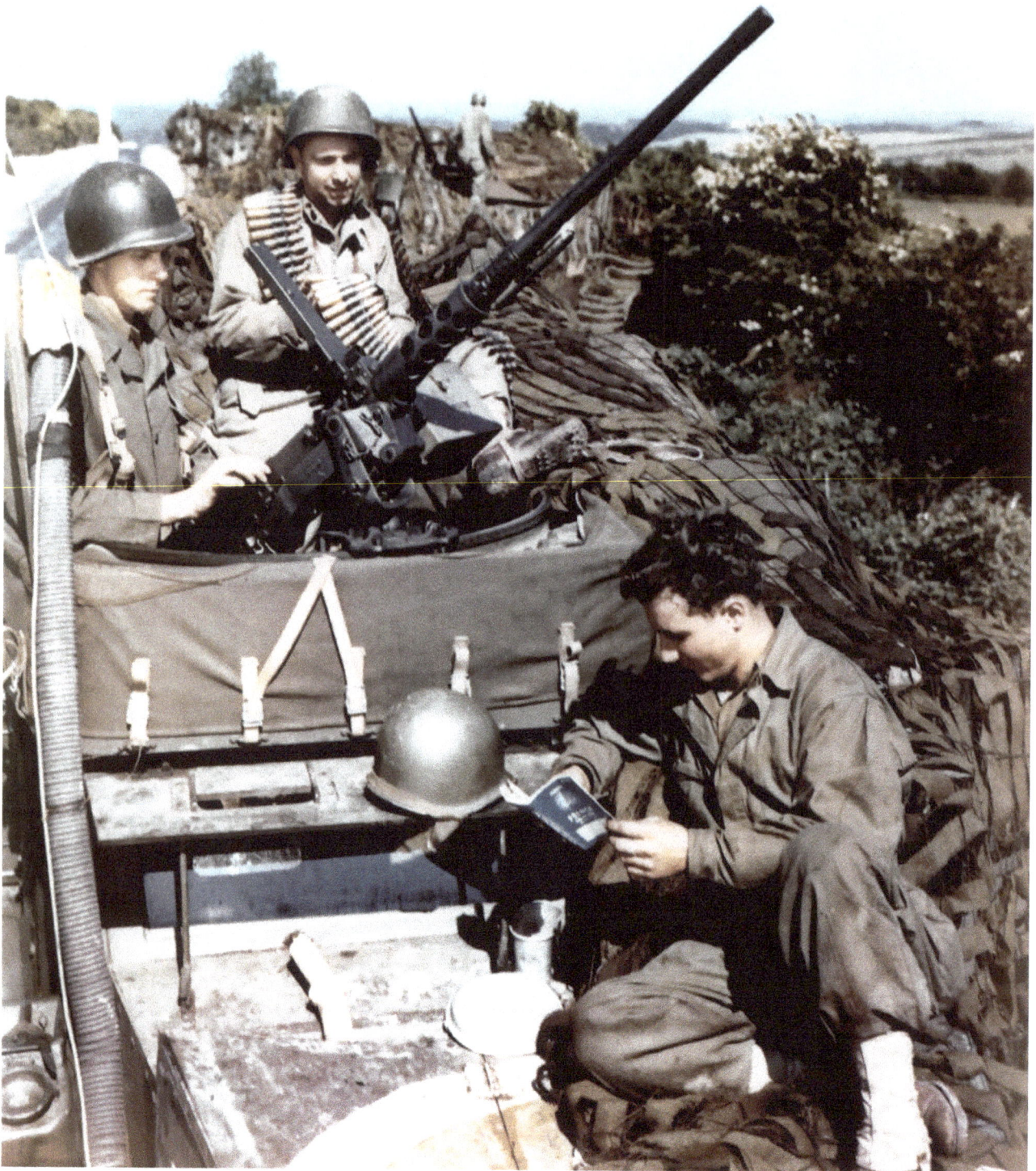

The crew of the half-track above in England are studying the small booklet on France in preparation for D-day in Normandy. Note the details of the .50 caliber machine gun mount and the deep water fording gear.

The half-track 40mm gun motor carriage M34 above is on duty at the Taegu Air Force Base in Korea. It is part of the 76th Antiaircraft Battalion in I Corps. This photograph was dated 13 January 1951

ACCEPTANCES OF HALF-TRACK VEHICLES FROM NEW PRODUCTION

Vehicle	Total Acceptances	First Acceptances	Final Acceptances
Half-Track Car M2	11,415	May 1941	September 1943
Half-Track Car M2A1	1643	October 1943	March 1944
Half-Track Personnel Carrier M3	12,391	May 1941	September 1943
Half-Track Personnel Carrier M3A1*	4222	October 1943	June 1945
81mm Mortar Carrier M4	572	August 1941	October 1942
81mm Mortar Carrier M4A1	600	May 1943	October 1943
81mm Mortar Carrier M21	110	January 1944	March 1944
Half-Track Personnel Carrier M5	4625	December 1942	September 1943
Half-Track Personnel Carrier M5A1	2959	October 1943	March 1944
Half-Track Car M9A1	3433	March 1943	December 1943
75mm Gun Motor Carriage M3**	2202	August 1941	April 1943
57mm Gun Motor Carriage T48	962	December 1942	May 1943
105mm Howitzer Motor Carriage T19	324	January 1942	April 1942
75mm Howitzer Motor Carriage T30	500	February 1942	November 1942
Multiple Gun Motor Carriage T28E1	80	July 1942	August 1942
Multiple Gun Motor Carriage M15	600	February 1943	April 1943
Combination Gun Motor Carriage M15A1	1652	October 1943	February 1944
Multiple Gun Motor Carriage M13	1103	January 1943	May 1943
Multiple Gun Motor Carriage M14	1605	December 1942	December 1943
Multiple Gun Motor Carriage M16 ***	3614	May 1943	November 1944
Multiple Gun Motor Carriage M17	1000	December 1943	March 1944
Twin 20mm Gun Motor Carriage T10E1	110	March 1944	March 1944

* Includes 1360 vehicles converted from the 75mm gun motor carriage M3
** Includes 86 75mm gun motor carriages T12
*** Includes 737 vehicles converted from the twin 20mm gun motor carriage T10E1 (109) and the multiple gun motor carriage M13 (628)

ACCEPTANCES OF REMANUFACTURED HALF-TRACKS

Vehicle	Total Acceptances	First Acceptances	Final Acceptances
Half-Track Car M2A1 (from M2)	1266	January 1944	June 1945
Half-Track Personnel Carrier M3 (from T30)	108	August 1942	August 1942
Half-Track Personnel Carrier M3A1 (from T48)	281	January 1944	May 1944
Half-Track Personnel Carrier M3A1 (from M3)	2209	May 1944	June 1945
Half-Track Personnel Carrier M3A1 (from T19)	90	July 1945	July 1945
Half-Track Personnel Carrier M3A1 (from M13)	1	July 1945	July 1945
Half-Track Personnel Carrier M3A1 (from T30)	1	June 1945	June 1945
Half-Track Personnel Carrier M5	3209	July 1944	June 1945
Half-Track Personnel Carrier M5A1	65	March 1945	April 1945
Half-Track Car M9A1	791	April 1945	June 1945

A major problem in the selection of material for inclusion in these data sheets was frequently not a lack of information, but the large number of sources available, many of which did not agree. These sources included the original design values from specifications and production drawings, measurements and test results from the evaluation of the vehicle at the proving ground, and the final results after tests by the user such as the Armored Board. Many of these differences resulted from changes in stowage and personnel during the development and service life of the vehicle.

Data sheets are included for all of the production half-tracks as well as a number of the experimental vehicles. An effort has been made to obtain comparable values for all of the vehicles. Dimensions have been taken from the production drawings when these were available. When these were not available, test reports from Aberdeen Proving Ground and the Armored Board at Fort Knox were used. Obviously some measurements such as height and ground clearance varied with changes in the spring compression under differing loads on the vehicle. The long service of many of the half-track vehicles resulted in numerous stowage changes. When possible, stowage and the appropriate weight for the period of the vehicles greatest employment have been used in the data sheets. For production vehicles, the weight is usually rounded off to the nearest 1000 pounds and the combat weights included the crew and a full load of fuel and ammunition. If available, the exact weight of an experimental vehicle is used, although in some cases, only estimated values could be obtained.

Most of the terms in the data sheets are self-explanatory, but some may need clarification. For example, the term fire height is included on the sheets for some of the self-propelled weapons. This is the distance from the ground to the center line of the main weapon bore at zero elevation. The tread is the distance between the center line of the wheels or tracks. The wheelbase for a half-track vehicle is the distance between the center of the front wheels to the center of the track suspension. The terms left or right are in reference to someone sitting in the vehicle driver's seat. The engine horsepower and torque are reported as net values including the effect of all accessories installed in the vehicle. All of the power to weight ratios were calculated using the combat weight of the vehicle. The angles of approach and departure are shown in the following sketch.

The armor in the data sheets is specified by the type, thickness, and angle with the vertical. This angle is shown in the sketch as the angle alpha between a

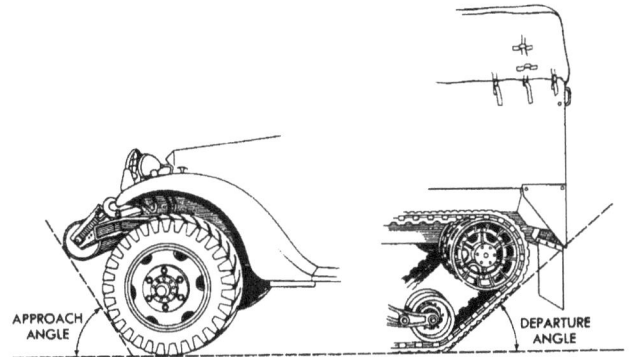

vertical plane and the surface of the armor plate. Also note that in this two dimensional sketch, alpha is equal to the angle beta, the angle of obliquity. The latter is defined as the angle between a line perpendicular to the plate surface and the path of a projectile impacting the armor. This angle is used to specify the armor penetration performance for various types of projectiles.

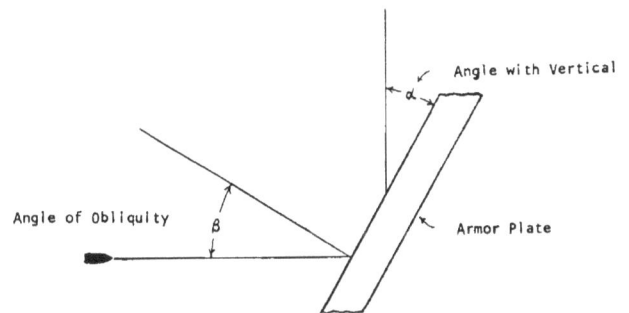

Face-hardened steel armor was installed on all of the half-tracks manufactured by the White Motor Company, the Autocar Company, and the Diamond T Motor Car Company. This armor was $1/4$ inch thick on all surfaces except for the windshield cover which was $1/2$ inch thick. The sides and rear surfaces were vertical. The later vehicles produced by the International Harvester Company were fitted with welded homogeneous steel armor $5/16$ inches thick except for the windshield cover which was increased to $5/8$ inches thick. Although the homogeneous armor was less vulnerable to the penetration of bullet splash and the dislocation of the cap screws, it did not offer the protection of the face-hardened plate. The $1/4$ inch thick face-hardened plate could be penetrated by the .30 caliber M2 armor piercing round up to a range of about 300 yards while the same round could penetrate the $5/16$ inch thick homogeneous steel armor up to almost 500 yards.

HALF-TRACK CARS M2 AND M2A1

GENERAL DATA

Crew:	10 men
Length: w/roller	234.75 inches
w/winch	241.63 inches
Width: Over side armor	77.25 inches
Over mine racks	87.50 inches
Height: Overall, M2	89 inches
M2A1	106 inches
Tread: Front	64.5 inches
Rear	63.8 inches
Wheelbase:	135.5 inches
Ground Clearance:	11.2 inches
Approach Angle: w/roller	37 degrees
w/winch	32 degrees
Departure Angle:	35 degrees
Weight, Combat Loaded: M2	19,195 pounds
M2A1	19,600 pounds
Power to Weight Ratio: Net, M2	15.3 hp/ton
M2A1	15.0 hp/ton
Winch Capacity:	10,000 pounds

ARMOR

Type: Rolled face hardened steel; Bolted assembly

Thickness:	Actual	Angle w/Vertical
Front, Radiator louvers	0.25 inches (6.4mm)	26 degrees
Windshield cover	0.50 inches (12.7mm)	25 degrees
Sides	0.25 inches (6.4mm)	0 degrees
Rear	0.25 inches (6.4mm)	0 degrees
Top, Hood only	0.25 inches (6.4mm)	83 degrees

ARMAMENT

(1) .50 caliber MG HB M2 flexible on skate mount (M2)
(1) .50 caliber MG HB M2 flexible on M49 ring mount (M2A1)
(1) .30 caliber MG M1919A4 on skate mount (M2)
(1) .30 caliber MG M1919A4 on pintle mount (M2A1)
Provision for (1) .45 caliber SMG M3 or M1928A1

AMMUNITION

700 rounds .50 caliber
7750 rounds .30 caliber
540 rounds .45 caliber
10 hand grenades
14 antitank mines M1A1

VISION EQUIPMENT

Vision slots in windshield armor and front door armor
Open top vehicle

ENGINE

Make and Model: White 160AX	
Type: 6 cylinder, 4 cycle, in-line	
Cooling System: Liquid	Ignition: Battery
Displacement:	386 cubic inches
Bore and Stroke:	4 x 5.125 inches
Compression Ratio:	6.44:1
Net Horsepower: (max)	147 hp at 3000 rpm
Net Torque: (max)	325 ft-lbs at 1200 rpm
Weight:	1015 pounds, dry
Fuel: 72 octane gasoline	60 gallons
Engine Oil:	12 quarts

POWER TRAIN

Master Clutch: Dry, single plate
Transmission: Constant mesh

Gear Ratios:	1st	4.92:1	4th	1.00:1
	2nd	2.60:1	reverse	4.37:1
	3rd	1.74:1		

Transfer Case: Constant mesh, direct and underdrive
  Gear Ratios: high 1.00:1   low 2.48:1
Differential: Front axle
  Gear Ratio: 6.8:1
Differential: Track drive
  Gear Ratio: 4.44:1
Drive Sprocket: At front of track with 18 teeth
  Pitch Diameter: 22.918 inches

RUNNING GEAR

Suspension:
  Front: Semi-elliptic longitudinal leaf spring
    2 ventilated disc steel wheels (1/side)
    Combat tires, 12 ply
    Tire Size: 8.25 x 20
  Rear: Vertical volute spring
    8 dual wheels in two bogies (1 bogie/track)
    Tire Size: 12 x 3.75 inches (or 12 x 4.125)
    2 dual track return rollers (1/track)
    Adjustable fixed idler at rear of each track (early)
    Spring loaded idler at rear of each track (late)
    Idler Size: 18.5 x 9.375 inches
    Track Type: Endless band
    Track Width: 12 inches
    Track Pitch: 4 inches, 58 pitches/track
    Track Ground Contact Length: 46.75 inches

ELECTRICAL SYSTEM

Nominal Voltage: 12 volts DC
Main Generator: (1) 12 volts, 55 amperes, driven by main engine
Auxiliary Generator: None
Battery: (1) 12 volts

COMMUNICATIONS

Radio: SCR 193 or 506 and 508 and 593; SCR 284 and 508 and 593;
    SCR 193 or 506 and 508 or 528 or 608 or 610 or 628.

FIRE AND GAS PROTECTION

(1) 2 pound carbon dioxide, portable
(2) 1½ quart decontaminating apparatus M2

PERFORMANCE

Maximum Speed: Level road	45 miles/hour
Maximum Grade:	60 per cent
Maximum Vertical Wall:	12 inches
Maximum Fording Depth:	32 inches
Minimum Turning Circle: (diameter)	59 feet
Cruising Range: Roads	approx. 200 miles

GENERAL DATA

Crew: M3 and M3A1	13 men
M3A2	5 to 12 men
Length: w/roller	242.63 inches
w/winch	249.63 inches
Width: Over side armor	77.25 inches
Over mine racks	87.50 inches
Height: Overall, M3	89 inches
M3A1 and M3A2	106 inches
Tread: Front	64.5 inches
Rear	63.8 inches
Wheelbase:	135.5 inches
Ground Clearance:	11.2 inches
Approach Angle: w/roller	37 degrees
w/winch	32 degrees
Departure Angle:	35 degrees
Weight, Combat Loaded: M3	20,000 pounds
M3A1	20,500 pounds
M3A2	21,200 pounds
Power to Weight Ratio: Net, M3	14.7 hp/ton
M3A1	14.3 hp/ton
M3A2	13.9 hp/ton
Winch Capacity:	10,000 pounds

ARMOR

Type: Rolled face hardened steel; Bolted assembly

Thickness:	Actual	Angle w/Vertical
Front, Radiator louvers	0.25 inches (6.4 mm)	26 degrees
Windshield cover	0.50 inches (12.7mm)	25 degrees
Sides	0.25 inches (6.4mm)	0 degrees
Rear	0.25 inches (6.4mm)	0 degrees
Top, Hood only	0.25 inches (6.4mm)	83 degrees

ARMAMENT

(1) .50 caliber MG HB M2 flexible on M49 ring mount (M3A1, M3A2)

(1) .30 caliber MG M1919A4 on pedestal mount M25 (M3)

(1) .30 caliber MG M1919A4 on pintle mount (M3A1, M3A2)

Provision for (1) .45 caliber SMG M3 or M1928A1

Provision for (12) .30 caliber Rifles M1 or Carbines M1

Provision for (1) 2.36 inch Rocket Launcher M1A1 or M9 (M3A2)

AMMUNITION

700 rounds .50 caliber (M3A1)

330 rounds .50 caliber (M3A2)*

4000 rounds .30 caliber (M3)

7750 rounds .30 caliber (M3A1)

2000 rounds .30 caliber (M3A2)*

540 rounds .45 caliber (M3, M3A1)

180 rounds .45 caliber (M3A2)

22 hand grenades (M3, M3A1)

24 hand grenades (M3A2)

6 2.36 inch antitank rockets M6 (M3A2)

24 antitank mines M1A1

VISION EQUIPMENT

Vision slots in windshield armor and front door armor

Open top vehicle

* When required, 600 additional rounds of .50 caliber or 6000 additional rounds of .30 caliber ammunition could be carried with the personnel capacity reduced by two man.

ENGINE

Make and Model: White 160AX	
Type: 6 cylinder, 4 cycle, in-line	
Cooling System: Liquid	Ignition: Battery
Displacement:	386 cubic inches
Bore and Stroke:	4 x 5.125 inches
Compression Ratio:	6.44:1
Net Horsepower: (max)	147 hp at 3000 rpm
Net Torque: (max)	325 ft-lbs at 1200 rpm
Weight:	1015 pounds, dry
Fuel: 72 octane gasoline	60 gallons
Engine Oil:	12 quarts

POWER TRAIN

Master Clutch: Dry, single plate

Transmission: Constant mesh

Gear Ratios:	1st	4.92:1	4th	1.00:1
	2nd	2.60:1	reverse	4.37:1
	3rd	1.74:1		

Transfer Case: Constant mesh, direct and underdrive

Gear Ratios: high 1.00:1   low 2.48:1

Differential: Front axle

Gear Ratio: 6.8:1

Differential: Track drive

Gear Ratio: 4.44:1

Drive Sprocket: At front of track with 18 teeth

Pitch Diameter: 22.918 inches

RUNNING GEAR

Suspension:

Front: Semi-elliptic longitudinal leaf spring

2 ventilated disc steel wheels (1/side)

Combat tires, 12 ply

Tire Size: 8.25 x 20

Rear: Vertical volute spring

8 dual wheels in two bogies (1 bogie/track)

Tire Size: 12 x 3.75 inches (or 12 x 4.125)

2 dual track return rollers (1/track)

Adjustable fixed idler at rear of each track (early)

Spring loaded idler at rear of each track (late)

Idler Size: 18.5 x 9.375 inches

Track Type: Endless band

Track Width: 12 inches

Track Pitch: 4 inches, 58 pitches/track

Track Ground Contact Length: 46.75 inches

ELECTRICAL SYSTEM

Nominal Voltage: 12 volts DC

Main Generator: (1) 12 volts, 55 amperes, driven by main engine

Auxiliary Generator: None

Battery: (1) 12 volts

COMMUNICATIONS

Radio: SCR 193 or 506 and 508 and 593; SCR 284 and 508 and 593; SCR 193 or 506 and 508 or 528 or 608 or 610 or 628.

FIRE AND GAS PROTECTION

(1) 2 pound carbon dioxide, portable

(2) 1½ quart decontaminating apparatus M2

PERFORMANCE

Maximum Speed: Level road	45 miles/hour
Maximum Grade:	60 per cent
Maximum Vertical Wall:	12 inches
Maximum Fording Depth:	32 inches
Minimum Turning Circle: (diameter)	59 feet
Cruising Range: Roads approx.	200 miles

HALF-TRACK CAR M9A1

GENERAL DATA

Crew:	10 men
Length: w/roller	242.19 inches
w/winch	249.06 inches
Width: Over mine racks	86.875 inches
Height: Over .50 cal. MG	108 inches
Tread: Front	66.5 inches
Rear	63.8 inches
Wheelbase:	135.5 inches
Ground Clearance:	11.2 inches
Approach Angle: w/roller	40 degrees
w/winch	36 degrees
Departure Angle:	32 degrees
Weight, Combat Loaded:	21,200 pounds
Power to Weight Ratio: Net	13.5 hp/ton
Winch Capacity:	10,000 pounds

ARMOR

Type: Rolled homogeneous steel; Welded assembly

Thickness:	Actual	Angle w/Vertical
Front, Radiator louvers	0.31 inches (7.9mm)	27 degrees
Windshield cover	0.625 inches (15,9mm)	23 degrees
Sides	0.31 inches (7.9mm)	0 degrees
Rear	0.31 inches (7.9mm)	0 degrees
Top, Hood only	0.31 inches (7.9mm)	83 degrees

ARMAMENT

(1) .50 caliber MG HB M2 flexible on M49 ring mount
(1) .30 caliber MG M1919A4 on pintle mount
Provision for (1) .45 caliber SMG M3 or M1928A1

AMMUNITION

700 rounds .50 caliber
7750 rounds .30 caliber
540 rounds .45 caliber
10 hand grenades
14 antitank mines M1A1

VISION EQUIPMENT

Vision slots in windshield armor and front door armor
Open top vehicle

ENGINE

Make and Model: International Harvester RED-450-B
Type: 6 cylinder, 4 cycle, in-line
Cooling System: Liquid   Ignition: Battery

Displacement:	450.99 cubic inches
Bore and Stroke:	4.375 x 5 inches
Compression Ratio:	6.35:1
Net Horsepower: (max)	143 hp at 2700 rpm
Net Torque: (max)	348 ft-lbs at 800 rpm
Weight:	1250 pounds, dry
Fuel: 72 octane gasoline	60 gallons
Engine Oil:	10.5 quarts

POWER TRAIN

Master Clutch: Dry, single disc
Transmission: Constant mesh

Gear Ratios:	1st	4,92:1	4th	1.00:1
	2nd	2.60:1	reverse	4.37:1
	3rd	1.74:1		

Transfer Case: Constant mesh, direct and underdrive
Gear Ratios: high 1.00:1   low 2.48:1
Differential: Front axle:
Gear Ratio: 7.16:1
Differential: Track drive
Gear Ratio: 4.22:1
Drive Sprocket: At front of track with 18 teeth
Pitch Diameter: 22.918 inches

RUNNING GEAR

Suspension:
Front: Semi-elliptic longitudinal leaf spring
2 ventilated disc steel wheels (1/side)
Combat tires, 12 ply
Tire Size: 9.00 x 20
Rear: Vertical volute spring
8 dual wheels in two bogies (1 bogie/track)
Tire Size: 12 x 3.75 inches (or 12 x 4.125)
2 dual track return rollers (1/track)
Spring loaded idler at rear of each track
Idler Size: 18.5 x 9.375 inches
Track Type: Endless band
Track Width: 12 inches
Track Pitch: 4 inches, 58 pitches/track
Track Ground Contact Length: 46.75 inches

ELECTRICAL SYSTEM

Nominal Voltage: 12 volts DC
Main Generator: (1) 12 volts, 55 amperes, driven by main engine
Auxiliary Generator: None
Battery: (1) 12 volts

COMMUNICATIONS

Radio: SCR 193 or 506 and 508 and 593; SCR 284 and 508 and 593;
SCR 193 or 506 and 508 or 528 or 608 or 610 or 628.

FIRE AND GAS PROTECTION

(1) 2 pound carbon dioxide, portable
(2) 1½ quart decontaminating apparatus M2

PERFORMANCE

Maximum Speed: Level road	42 miles/hour
Maximum Grade:	60 per cent
Maximum Vertical Wall:	12 inches
Maximum Fording Depth:	32 inches
Minimum Turning Circle: (diameter)	59 feet
Cruising Range: Roads	approx. 200 miles

GENERAL DATA

Crew: M5 and M5A1	13 men
M5A2	5 to 12 men
Length: w/roller	242.19 inches
w/winch	249.06 inches
Width: Over mine racks	86.875 inches
Height: Over .50 cal. MG	108 inches
Tread: Front	66.5 inches
Rear	63.8 inches
Wheelbase:	135.5 inches
Ground Clearance:	11.2 inches
Approach Angle: w/roller	40 degrees
w/winch	36 degrees
Departure Angle:	32 degrees
Weight, Combat Loaded: M5	20,500 pounds
M5A1	21,500 pounds
M5A2	22,500 pounds
Power to Weight Ratio: Net, M5	14.0 hp/ton
M5A1	13.3 hp/ton
M5A2	12.7 hp/ton
Winch Capacity:	10,000 pounds

ARMOR

Type: Rolled homogeneous steel; Welded assembly

Thickness:	Actual	Angle w/Vertical
Front, Radiator louvers	0.31 inches (7.9mm)	27 degrees
Windshield cover	0.625 inches (15.9mm)	23 degrees
Sides	0.31 inches (7.9mm)	0 degrees
Rear	0.31 inches (7.9mm)	0 degrees
Top, Hood only	0.31 inches (7.9mm)	83 degrees

ARMAMENT

(1) .50 caliber MG HB M2 flexible on M49 ring mount (M5A1, M5A2)
(1) .30 caliber MG M1919A4 on pedestal mount M25 (M5)
(1) .30 caliber MG M1919A4 on pintle mount (M5A1, M5A2)
Provision for (1) .45 caliber SMG M3 or M1928A1
Provision for (12) .30 caliber Rifles M1 or Carbines M1
Provision for (1) 2.36 inch Rocket Launcher M1A1 or M9 (M5A2)

AMMUNITION

700 rounds .50 caliber (M5A1)
330 rounds .50 caliber (M5A2)*
4000 rounds .30 caliber (M5)
7750 rounds .30 caliber (M5A1)
2000 rounds .30 caliber (M5A2)*
540 rounds .45 caliber (M5, M5A1)
180 rounds .45 caliber (M5A2)
22 hand grenades (M5, M5A1)
24 hand grenades (M5A2)
6 2.36 inch antitank rockets M6 (M5A2)
24 antitank mines M1A1

VISION EQUIPMENT

Vision slots in windshield armor and front door armor
Open top vehicle

* When required, 600 additional rounds of .50 caliber or 6000 additional rounds of .30 caliber ammunition could be carried with the personnel capacity reduced by two men.

ENGINE

Make and Model: International Harvester RED-450-B	
Type: 6 cylinder, 4 cycle, in-line	
Cooling System: Liquid	Ignition: Battery
Displacement:	450.99 cubic inches
Bore and Stroke:	4.375 x 5 inches
Compression Ratio:	6.35:1
Net Horsepower: (max)	43 hp at 2700 rpm
Net Torque: (max)	348 ft-lbs at 800 rpm
Weight:	1250 pounds, dry
Fuel: 72 octane gasoline	60 gallons
Engine Oil:	10.5 quarts

POWER TRAIN

Master Clutch: Dry, single disc
Transmission: Constant mesh

Gear Ratios:			
1st	4.92:1	4th	1.00:1
2nd	2.60:1	reverse	4.37:1
3rd	1.74:1		

Transfer Case: Constant mesh, direct and underdrive
Gear Ratios: high 1.00:1 low 2.48:1
Differential: Front axle
Gear Ratio: 7.16:1
Differential: Track drive
Gear Ratio: 4.22:1
Drive Sprocket: At front of track with 18 teeth
Pitch Diameter: 22.918 inches

RUNNING GEAR

Suspension:
Front: Semi-elliptic longitudinal leaf spring
2 ventilated disc steel wheels (1/side)
Combat tires, 12 ply
Tire Size: 9.00 x 20
Rear: Vertical volute spring
8 dual wheels in two bogies (1 bogie/track)
Tire Size: 12 x 3.75 inches (or 12 x 4.125)
2 dual track return rollers (1/track)
Spring loaded idler at rear of each track
Idler Size: 18.5 x 9.375 inches
Track Type: Endless band
Track Width: 12 inches
Track Pitch: 4 inches, 58 pitches/track
Track Ground Contact Length: 46.75 inches

ELECTRICAL SYSTEM

Nominal Voltage: 12 volts DC
Main Generator: (1) 12 volts, 55 amperes, driven by main engine
Auxiliary Generator: None
Battery: (1) 12 volts

COMMUNICATIONS

Radio: SCR 193 or 506 and 508 and 593; SCR 284 and 508 and 593; SCR 193 or 506 and 508 or 528 or 608 or 610 or 628.

FIRE AND GAS PROTECTION

(1) 2 pound carbon dioxide, portable
(2) 1½ quart decontaminating apparatus M2

PERFORMANCE

Maximum Speed: Level road		42 miles/hour
Maximum Grade:		60 per cent
Maximum Vertical Wall:		12 inches
Maximum Fording Depth:		32 inches
Minimum Turning Circle: (diameter)		59 feet
Cruising Range: Roads	approx.	200 miles

81mm MORTAR CARRIERS M4 AND M4A1
(Based upon the half-track car M2)

GENERAL DATA

Crew:	6 men
Length: w/roller	236.75 inches
w/winch	243.625 inches
Width: Over side armor	77.25 inches
Height:	89.375 inches
Tread: Front:	64.5 inches
Rear	63.8 inches
Wheelbase:	135.5 inches
Ground Clearance:	11.2 inches
Approach Angle: w/roller	37 degrees
w/winch	32 degrees
Departure Angle:	35 degrees
Weight, Combat Loaded: M4	17,350 pounds
M4A1	18,000 pounds
Power to weight Ratio: M4	16.9 hp/ton
M4A1	16.3 hp/ton
Winch Capacity:	10,000 pounds

ARMOR

Type: Rolled face hardened steel; Bolted assembly

Thickness:	Actual	Angle w/Vertical
Front, Radiator louvers	0.25 inches (6.4mm)	26 degrees
Windshield cover	0.50 inches (12.7mm)	25 degrees
Sides	0.25 inches (6.4mm)	0 degrees
Rear	0.25 inches (6.4mm)	0 degrees
Top, Hood only	0.25 inches (6.4mm)	83 degrees

ARMAMENT

Primary: 81mm mortar M1 on mortar mount M1 aimed to rear

Traverse: Manual, M4	130 mils (65 left, 65 right)
M4A1	600 mils
Elevation: Manual	+80 to +40 degrees
Firing Rate: (max)	30 to 35 rounds/minute
Loading System:	Manual
Stabilizer System:	None

Secondary:
(1) .30 caliber MG M1919A4 on skate mount*
Provision for (1) .45 caliber SMG M3
Provision for (1) 2.36 inch rocket launcher M1A1

AMMUNITION

96 rounds 81mm
540 rounds .45 caliber
2000 rounds .30 caliber
10 2.36 inch antitank rockets M6
10 hand grenades
14 antitank mines M1A1

FIRE CONTROL AND VISION EQUIPMENT

Primary Weapon: Sight M6
Vision Devices:
Vision slots in windshield armor and front door armor
Open top vehicle

*Some early vehicles armed with .30 caliber MG M1917A1
and a .50 caliber MG HB M2 carried on some vehicles.

ENGINE

Make and Model: White 160AX	
Type: 6 cylinder, 4 cycle, in-line	
Cooling System: Liquid	Ignition: Battery
Displacement:	386 cubic inches
Bore and Stroke:	4 x 5.125 inches
Compression Ratio:	6.44:1
Net Horsepower: (max)	147 hp at 3000 rpm
Net Torque: (max)	325 ft-lbs at 1200 rpm
Weight:	1015 pounds, dry
Fuel: 72 octane gasoline	60 gallons
Engine Oil:	12 quarts

POWER TRAIN

Master Clutch: Dry, single plate
Transmission: Constant mesh

Gear Ratios:	1st	4.94:1	4th	1.00:1
	2nd	2.60:1	reverse	4.37:1
	3rd	1.74:1		

Transfer Case: Constant mesh, direct and underdrive
Gear Ratios: high 1.00:1 low 2.48:1
Differential: Front axle
Gear Ratio: 6.8:1
Differential: Track drive
Gear Ratio: 4.44:1
Drive Sprocket: At front of track with 18 teeth
Pitch Diameter: 22.918 inches

RUNNING GEAR

Suspension:
Front: Semi-elliptic longitudinal leaf spring
2 ventilated disc steel wheels (1/side)
Combat tires, 12 ply
Tire Size: 8.25 x 20
Rear: Vertical volute spring
8 dual wheels in two bogies (1bogie/track)
Tire Size: 12 x 3.75 inches (or 12 x 4.125)
2 dual track return rollers (1/track)
Adjustable fixed idler at rear of each track (early)
Spring loaded idler at rear of each track (late)
Idler Size: 18.5 x 9.375 inches
Track Type: Endless band
Track Width: 12 inches
Track Pitch: 4 inches, 58 pitches/track
Track Ground Contact Length: 46.75 inches

ELECTRICAL SYSTEM

Nominal Voltage: 12 volts DC
Main Generator: (1) 12 volts, 55 amperes, driven by main engine
Auxiliary Generator: None
Battery: (1) 12 volts

COMMUNICATIONS

Radio: SCR 509 or SCR 510
Flag Set M238

FIRE AND GAS PROTECTION

(1) 2 pound carbon dioxide, portable
(3) 1½ quart decontaminating apparatus M2

PERFORMANCE

Maximum Speed: Level road		45 miles/hour
Maximum Grade:		60 per cent
Maximum Vertical Wall:		12 inches
Maximum Fording Depth:		32 inches
Minimum Turning Circle: (diameter)		59 feet
Cruising Range: Roads	approx.	200 miles

81mm MORTAR CARRIER M21
(Based upon the half track personnel carrier M3)

GENERAL DATA

Crew:	6 men
Length: w/winch	244.875 inches
Width: Over mine racks	87.5 inches
Height:	89 inches
Tread: Front	64.5 inches
Rear	63.8 inches
Wheelbase:	135.5 inches
Ground Clearance:	11.2 inches
Approach Angle: w/winch	32 degrees
Departure Angle:	35 degrees
Weight, Combat Loaded:	20,000 pounds
Power to Weight Ratio: Net	14.7 hp/ton
Winch Capacity:	10,000 pounds

ARMOR

Type: Rolled face hardened steel; Bolted assembly

Thickness:	Actual	Angle w/Vertical
Front, Radiator louvers	0.25 inches (6.4mm)	26 degrees
Windshield cover	0.50 inches (12.7mm)	25 degrees
Sides	0.25 inches (6.4mm)	0 degrees
Rear	0.25 inches (6.4mm)	0 degrees
Top, Hood only	0.25 inches (6.4mm)	83 degrees

ARMAMENT

Primary: 81mm mortar M1 on mortar mount M1 aimed to front

Traverse: Manual	60 degrees (30 left, 30 right)
Elevation: Manual	+85 to +40 degrees
Firing Rate: (max)	30 to 35 rounds/minute
Loading System:	Manual
Stabilizer System:	None

Secondary:

(1) .50 caliber MG HB M2 flexible on pedestal mount
Provision for (1) .45 caliber SMG M3
Provision for (1) 2.36 inch rocket launcher M1A1

AMMUNITION

97 rounds 81mm
400 rounds .50 caliber
600 rounds .45 caliber
6 2.36 inch antitank rockets M6
12 hand grenades
12 antitank mines M1A1

FIRE CONTROL AND VISION EQUIPMENT

Primary Weapon: Sight M6
Vision Devices:
Vision slots in windshield armor and front door armor
Open top vehicle

ENGINE

Make and Model: White 160AX	
Type: 6 cylinder, 4 cycle, in-line	
Cooling System: Liquid	Ignition: Battery
Displacement:	386 cubic inches
Bore and Stroke:	4 x 5.125 inches
Compression Ratio:	6.44:1
Net Horsepower: (max)	147 hp at 3000 rpm
Net Torque: (max)	325 ft-lbs at 1200 rpm
Weight:	1015 pounds, dry
Fuel: 72 octane gasoline	60 gallons
Engine Oil:	12 quarts

POWER TRAIN

Master Clutch: Dry, single plate
Transmission: Constant mesh

Gear Ratios:	1st	4.92:1	4th	1.00·1
	2nd	2.60:1	reverse	4.37:1
	3rd	1.74:1		

Transfer Case: Constant mesh, direct and underdrive
Gear Ratio: high: 1.00:1   low 2.48:1
Differential: Front axle
Gear Ratio: 6.8:1
Differential: Track drive
Gear Ratio: 4.44:1
Drive Sprocket: At front of track with 18 teeth
Pitch Diameter: 22.918 inches

RUNNING GEAR

Suspension:
Front: Semi-elliptic longitudinal leaf spring
2 ventilated disc steel wheels (1/side)
Combat tires, 12 ply
Tire Size: 8.25 x 20
Rear: Vertical volute spring
8 dual wheels in two bogies (1 bogie/track)
Tire Size: 12 x 3.75 inches (or 12 x 4.125)
2 dual track return rollers (1/track)
Spring loaded idler at rear of each track
Idler Size: 18.5 x 9.375 inches
Track Type: Endless band
Track Width: 12 inches
Track Pitch: 4 inches, 58 pitches/track
Track Ground Contact Length: 46.75 inches

ELECTRICAL SYSTEM

Nominal Voltage: 12 volts DC
Main Generator: (1) 12 volts, 55 amperes, driven by main engine
Auxiliary Generator: None
Battery: (1) 12 volts

COMMUNICATIONS

Radio: SCR 509 or SCR 510
Flag Set M238

FIRE AND GAS PROTECTION

(1) 2 pound carbon dioxide, portable
(3) 1½ quart decontaminating apparatus M2

PERFORMANCE

Maximum Speed: Level road	45 miles/hour
Maximum Grade:	60 per cent
Maximum Vertical Wall:	12 inches
Maximum Fording Depth:	32 inches
Minimum Turning Circle: (diameter)	59 feet
Cruising Range: Roads approx.	200 miles

4.2 inch MORTAR CARRIER T21E1
(Based upon the half-track personnel carrier M3)

GENERAL DATA

Crew:	5 men
Length: Over roller	236.75 inches
Width: Over side armor	77.25 inches
Height: Over .50 caliber MG	101 inches
Tread: Front	64.5 inches
Rear	63.8 inches
Wheelbase:	135.5 inches
Ground Clearance:	11.2 inches
Approach Angle: w/roller	37 degrees
Departure Angle:	35 degrees
Weight, Combat Loaded:	20,000 pounds
Power to Weight Ratio:	14.7 hp/ton

ARMOR

Type: Rolled face hardened steel; Bolted assembly

Thickness:	Actual	Angle w/Vertical
Front, Radiator louvers	0.25 inches (6.4mm)	26 degrees
Windshield cover	0.50 inches (12.7mm)	25 degrees
Sides	0.25 inches (6.4mm)	0 degrees
Rear	0.25 inches (6.4mm)	0 degrees
Top, hood only	0.25 inches (6.4mm)	83 degrees

ARMAMENT

Primary: 4.2 inch mortar M2 on mount aimed to front of vehicle

Traverse: Manual	11 degrees (200 mils)
Elevation: Manual	+ 60 to + 45 degrees
Firing Rate: (max)	3 rounds/minute
Loading System:	Manual
Stabilizer System:	None

Secondary:
> (1) .50 caliber MGHB M2 flexible on pedestal mount
> Provision for (1) .45 caliber SMG M3
> Provision for (1) 2.36 inch rocket launcher M1A1 or M9

AMMUNITION
> 90 rounds 4.2 inch
> 400 rounds .50 caliber
> 540 rounds .45 caliber
> 10 2.36 inch antitank rockets M6
> 12 hand grenades

FIRE CONTROL AND VISION EQUIPMENT

Primary Weapon: Sight M2
Vision Devices:
> Vision slots in windshield armor and front door armor
> Open top vehicle

ENGINE

Make and Model: White 160AX	
Type: 6 cylinder, 4 cycle, in-line	
Cooling System: Liquid	Ignition: Battery
Displacement:	386 cubic inches
Bore and Stroke:	4 x 5.125 inches
Compression Ratio:	6.44:1
Net Horsepower: (max)	147 hp at 3000rpm
Net Torque: (max)	325 ft-lbs at 1200 rpm
Weight:	1015 pounds, dry
Fuel: 72 octane gasoline	60 gallons
Engine Oil:	12 quarts

POWER TRAIN

Master Clutch: Dry, single plate
Transmission: Constant mesh

Gear Ratios:	1st	4.92:1	4th	1.00:1
	2nd	2.60:1	reverse	4.37:1
	3rd	1.74:1		

Transfer Case: Constant mesh, direct and underdrive

Gear Ratios:	high 1.00:1	low 2.48:1

Differential: Front axle
> Gear Ratio: 6.8:1
Differential: Track drive
> Gear Ratio: 4.44:1
Drive Sprocket: At front of track with 18 teeth
> Pitch Diameter: 22.918 inches

RUNNING GEAR

Suspension:
> Front: Semi-elliptic longitudinal leaf spring
> 2 ventilated disc steel wheels (1/side)
> Combat Tires, 12 ply
> Tire Size: 8.25 x 20
> Rear: Vertical volute spring
> 8 dual wheels in two bogies (1 bogie/track)
> Tire Size: 12 x 3.75 inches (or 12 x 4.125)
> 2 dual track return rollers (1/track)
> Spring loaded idler at rear of each track
> Idler Size: 18.5 x 9.375 inches
> Track Type: Endless band
> Track Width: 12 inches
> Track Pitch: 4 inches, 58 pitches/track
> Track Ground Contact Length: 46.75 inches

ELECTRICAL SYSTEM

Nominal Voltage: 12 volts DC
Main Generator: (1) 12 volts, 55 amperes, driven by main engine
Auxiliary Generator: None
Battery: (1) 12 volts

COMMUNICATIONS

Radio: SCR 509 or 510
Flag Set M238

FIRE AND GAS PROTECTION
> (1) 2 pound carbon dioxide, portable
> (3) 1½ quart decontaminating apparatus M2

PERFORMANCE

Maximum Speed: Level road		45 miles/hour
Maximum Grade:		60 per cent
Maximum Vertical Wall:		12 inches
Maximum Fording Depth:		32 inches
Minimum Turning Circle: (diameter)		59 feet
Cruising Range: Roads	approx.	200 miles

57mm GUN MOTOR CARRIAGE T48
(Based upon the half-track personnel carrier M3)

GENERAL DATA

Crew:	5 men
Length: w/roller	252.63 inches
Width: Over side armor	77.25 inches
Height: Over gun shield	90 inches
Fire Height:	72 inches
Tread: Front	64.5 inches
Rear	63.8 inches
Wheelbase:	135.5 inches
Ground Clearance:	11.2 inches
Approach Angle:	37 degrees
Departure Angle:	35 degrees
Weight, Combat Loaded:	19,000 pounds
Power to Weight Ratio:	15.5 hp/ton

ARMOR

Type: Rolled face hardened steel; Bolted assembly

Thickness:	Actual	Angle w/Vertical
Gun Shield, Front	0.625 inches (15.9mm)	26 degrees
Sides	0.25 inches (6.4mm)	0 degrees
Top	0.25 inches (6.4mm)	75 degrees
Front, Radiator louvers	0.25 inches (6.4mm)	26 degrees
Windshield cover	0.50 inches (12.7mm)	25 degrees
Sides	0.25 inches (6.4mm)	0 degrees
Rear	0.25 inches (6.4mm)	0 degrees
Top, Hood only	0.25 inches (6.4mm)	83 degrees

ARMAMENT

Primary: 57mm gun M1 on mount T5

Traverse: Manual	55 degrees
	(27.5 left, 27.5 right)
Elevation: Manual	+15 to -5 degrees
Firing Rate: (max)	30 rounds/minute
Loading System:	Manual
Stabilizer System:	None

Secondary:
  Provision for (5) .303 caliber Rifles, British
  Provision for (1) Grenade Launcher (for rifle)

AMMUNITION

  99 rounds 57mm
  10 rifle grenades M9A1
  12 hand grenades

FIRE CONTROL AND VISION EQUIPMENT

Primary Weapon:	Direct	Indirect
	Telescope M18	None

Vision Devices:
  Vision slots in windshield armor and front door armor
  Open top vehicle

ENGINE

Make and Model: White 160AX	
Type: 6 cylinder, 4 cycle, in-line	
Cooling System: Liquid	Ignition: Battery
Displacement:	386 cubic inches
Bore and Stroke:	4 x 5.125 inches
Compression Ratio:	6.44:1
Net Horsepower: (max)	147 hp at 3000 rpm
Net Torque: (max)	325 ft-lbs at 1200 rpm
Weight:	1015 pounds, dry
Fuel: 72 octane gasoline	60 gallons
Engine Oil:	12 quarts

POWER TRAIN

Master Clutch: Dry, single plate
Transmission: Constant mesh

Gear Ratios:	1st	4.92:1	4th	1.00:1
	2nd	2.60:1	reverse	4.37:1
	3rd	1.74:1		

Transfer Case: Constant mesh, direct and underdrive
  Gear Ratios: high 1.00:1   low 2.48:1
Differential: Front axle
  Gear Ratio: 6.8:1
Differential: Track drive
  Gear Ratio: 4.44:1
Drive Sprocket: At front of track with 18 teeth
  Pitch Diameter: 22.918 inches

RUNNING GEAR

Suspension:
  Front: Semi-elliptic longitudinal leaf spring
    2 ventilated disc steel wheels (1/side)
    Combat Tires: 12 ply
    Tire Size: 8.25 x 20
  Rear: Vertical volute spring
    8 dual wheels in two bogies (1 bogie/track)
    Tire Size: 12 x 3.75 inches (or 12 x 4.125)
    2 dual track return rollers (1/track)
    Adjustable fixed idler at rear of each track
    Idler Size: 18.5 x 9.375 inches
    Track Type: Endless band
    Track Width: 12 inches
    Track Pitch: 4 inches, 58 pitches/track
    Track Ground Contact Length: 46.75 inches

ELECTRICAL SYSTEM

Nominal Voltage: 12 volts DC
Main Generator: (1) 12 volts, 55 amperes, driven by main engine
Auxiliary Generator: None
Battery: (1) 12 volts

COMMUNICATIONS

Radio: Wireless Set No. 19, British
Flag Set M238

FIRE AND GAS PROTECTION

    (1) 2 pound carbon dioxide, portable
    (1) 1½ quart decontaminating apparatus M2

PERFORMANCE

Maximum Speed: Level road		45 miles/hour
Maximum Grade:		60 per cent
Maximum Vertical Wall:		12 inches
Maximum Fording Depth:		32 inches
Minimum Turning Circle: (diameter)		59 feet
Cruising Range: Roads	approx.	200 miles

75mm GUN MOTOR CARRIAGES M3 AND M3A1
(Based upon the half-track personnel carrier M3)

GENERAL DATA

Crew:	5 men
Length:	245.5 inches
Width: Over side armor	77.25 inches
Height: Over gunshield	98.6 inches
Fire Height:	81.9 inches
Tread: Front	64.5 inches
Rear	63.8 inches
Wheelbase:	135.5 inches
Ground Clearance:	11.2 inches
Approach Angle:	37 degrees
Departure Angle:	35 degrees
Weight, Combat Loaded:	20,000 pounds
Power to Weight Ratio: Net	14.7 hp/ton

ARMOR

Type: Rolled face hardened steel; Bolted assembly

Thickness:	Actual	Angle w/Vertical
Gun Shield, Front	0.625 inches (15.9mm)	12 degrees
Sides	0.25 inches (6.4mm)	0 degrees
Top	0.25 inches (6.4mm)	83 degrees
Front, Radiator louvers	0.25 inches (6.4mm)	26 degrees
Windshield cover	0.50 inches (12.7mm)	25 degrees
Sides	0.25 inches (6.4mm)	0 degrees
Rear	0.25 inches (6.4mm)	0 degrees
Top, Hood only	0.25 inches (6.4mm)	83 degrees

ARMAMENT

Primary: 75mm gun M1897A4 on mount M3 (M3) or M5 (M3A1)

Traverse: Manual, M3 mount		40 degrees
		(19 left, 21 right)
	M5 mount	42 degrees
		(21 left, 21 right)
Elevation: Manual, M3 mount		+29 to -10 degrees
	M5 mount	+29 to -6½ degrees
Firing Rate: (max)		20 rounds/minute
Loading System:		Manual
Stabilizer System:		None

Secondary:

Provision for (1) .30 caliber Rifle M1903
Provision for (4) .30 caliber Carbines M1
Provision for (1) Grenade Launcher (for rifle)

AMMUNITION

59 rounds 75mm
10 rifle grenades M9A1
12 hand grenades

FIRE CONTROL AND VISION EQUIPMENT

Primary Weapon:	Direct	Indirect
	Telescope M33	*

Vision Devices:

Vision slots in windshield armor and front door armor
Open top vehicle

* Indirect fire control equipment added at a later date.

ENGINE

Make and Model: White 160AX
Type: 6 cylinder, 4 cycle, in-line

Cooling System: Liquid	Ignition: Battery
Displacement:	386 cubic inches
Bore and Stroke:	4 x 5.125 inches
Compression Ratio:	6.44:1
Net Horsepower: (max)	147 hp at 3000 rpm
Net Torque: (max)	325 ft-lbs at 1200 rpm
Weight:	1015 pounds, dry
Fuel: 72 octane gasoline	60 gallons
Engine Oil:	12 quarts

POWER TRAIN

Master Clutch: Dry, single plate
Transmission: Constant mesh

Gear Ratios:				
1st	4.92:1	4th	1.00:1
2nd	2.60:1	reverse	4.37:1
3rd	1.74:1		

Transfer Case: Constant mesh, direct and underdrive

Gear Ratios: high 1.00:1 low 2.48:1

Differential: Front axle

Gear Ratio: 6.8:1

Differential: Track drive

Gear Ratio: 4.44:1

Drive Sprocket: At front of track with 18 teeth

Pitch Diameter: 22.918 inches

RUNNING GEAR

Suspension:

Front: Semi-elliptic longitudinal leaf spring
2 ventilated disc steel wheels (1/side)
Combat tires, 12 ply
Tire Size: 8.25 x 20

Rear: Vertical volute spring
8 dual wheels in two bogies (1bogie/track)
Tire Size: 12 x 3.75 inches (or 12 x 4.125)
2 dual track return rollers (1/track)
Adjustable fixed idler at rear of each track (early)
Spring loaded idler at rear of each track (late)
Idler Size: 18.5 x 9.375 inches
Track Type: Endless band
Track Width: 12 inches
Track Pitch: 4 inches, 58 pitches/track
Track Ground Contact Length: 46.75 inches

ELECTRICAL SYSTEM

Nominal Voltage: 12 volts DC
Main Generator: (1) 12 volts, 55 amperes, driven by main engine
Auxiliary Generator: None
Battery: (1) 12 volts

COMMUNICATIONS

Radio: SCR 510
Flag Set M238

FIRE AND GAS PROTECTION

(1) 2 pound carbon dioxide, portable
(1) 1½ quart decontaminating apparatus M2

PERFORMANCE

Maximum Speed: Level road	45 miles/hour
Maximum Grade:	60 per cent
Maximum Vertical Wall:	12 inches
Maximum Fording Depth:	32 inches
Minimum Turning Circle: (diameter)	59 feet
Cruising Range: Roads	approx. 200 miles

75mm HOWITZER MOTOR CARRIAGE T30
(Based upon the half-track personnel carrier M3)

GENERAL DATA

Crew:	5 men
Length:	240.1 inches
Width: Over side armor	77.25 inches
Height: Over howitzer shield	90 inches
Fire Height:	81.9 inches
Tread: Front	64.5 inches
Rear	63.8 inches
Wheelbase:	135.5 inches
Ground Clearance:	11.2 inches
Approach Angle:	37 degrees
Departure Angle:	35 degrees
Weight, Combat Loaded:	20,500 pounds
Power to Weight Ratio:	14.3 hp/ton

ARMOR

Type: Rolled face hardened steel; Bolted assembly

Thickness:	Actual	Angle w/Vertical
Howitzer Shield, Front	0.375 inches (9.5mm)	35 degrees
Sides	0.25 inches (6.4mm)	0 degrees
Front, Radiator louvers	0.25 inches (6.4mm)	26 degrees
Windshield cover	0.50 inches (12.7mm)	25 degrees
Sides	0.25 inches (6.4mm)	0 degrees
Rear	0.25 inches (6.4mm)	0 degrees
Top, Hood only	0.25 inches (6.4mm)	83 degrees

ARMAMENT

Primary: 75mm howitzer M1A1 on mount T10

Traverse: Manual	45 degrees
	(22.5 left, 22.5 right)
Elevation: Manual	+49.5 to -9 degrees
Firing Rate: (max)	8 rounds/minute
Loading System:	Manual
Stabilizer System:	None

Secondary:
> (1) .50 caliber MG HB M2 on pedestal mount M25, modified
> Provision for (1) .45 caliber SMG M1928A1
> Provision for (4) .30 caliber Carbines M1 or
>   (4) .30 caliber Rifles M1

AMMUNITION
> 60 rounds 75mm
> 300 rounds .50 caliber
> 200 rounds .45 caliber

FIRE CONTROL AND VISION EQUIPMENT

Primary Weapon:	Direct	Indirect
	Elbow Telescope M5	Panoramic Telescope M1
		Range Quadrant M3
		Gunner's Quadrant M1

Vision Devices:
> Vision slots in windshield armor and front door armor
> Open top vehicle

ENGINE

Make and Model: White 160AX	
Type: 6 cylinder, 4 cycle, in-line	
Cooling System: Liquid	Ignition: Battery
Displacement:	386 cubic inches
Bore and Stroke:	4 x 5.125 inches
Compression Ratio:	6.44:1
Net Horsepower: (max)	147 hp at 3000 rpm
Net Torque: (max)	325 ft-lbs at 1200 rpm
Weight:	1015 pounds, dry
Fuel: 72 octane gasoline	60 gallons
Engine Oil:	12 quarts

POWER TRAIN

Master Clutch: Dry, single plate

Transmission: Constant mesh

Gear Ratios:	1st	4.92:1	4th	1.00:1
	2nd	2.60:1	reverse	4.37:1
	3rd	1.74:1		

Transfer Case: Constant mesh, direct and underdrive

Gear Ratios:	high	1.00:1	low	2.48:1

Differential: Front axle
> Gear Ratio: 6.8:1

Differential: Track drive
> Gear Ratio: 4.44:1

Drive Sprocket: At front of track with 18 teeth
> Pitch Diameter: 22.918 inches

RUNNING GEAR

Suspension:
> Front: Semi-elliptic longitudinal leaf spring
>   2 ventilated disc steel wheels (1/side)
>   Combat tires, 12 ply
>   Tire Size: 8.25 x 20
> Rear: Vertical volute spring
>   8 dual wheels in two bogies (1 bogie/track)
>   Tire Size: 12 x 3.75 inches (or 12 x 4.125)
>   2 dual track return rollers (1/track)
>   Adjustable fixed idler at rear of each track
>   Idler Size: 18.5 x 9.375 inches
>   Track Type: Endless band
>   Track Width: 12 inches
>   Track Pitch: 4 inches, 58 pitches/track
>   Track Ground Contact Length: 46.75 inches

ELECTRICAL SYSTEM

Nominal Voltage: 12 volts DC

Main Generator: (1) 12 volts, 55 amperes, driven by main engine

Auxiliary Generator: None

Battery: (1) 12 volts

COMMUNICATIONS

Radio: SCR 510

Flag Set M238

FIRE PROTECTION
> (1) 1 quart carbon tetrachloride

PERFORMANCE

Maximum Speed: Level road	45 miles/hour
Maximum Grade:	60 per cent
Maximum Vertical Wall:	12 inches
Maximum Fording Depth:	32 inches
Minimum Turning Circle: (diameter)	59 feet
Cruising Range: Roads approx.	200 miles

105mm HOWITZER MOTOR CARRIAGE T19
(Based upon the half-track personnel carrier M3)

GENERAL DATA

Crew:	6 men
Length:	242.5 inches
Width: Over side armor	77.25 inches
Height:	92 inches
Fire Height:	79 inches
Tread: Front	64.5 inches
Rear	63.8 inches
Wheelbase:	135.5 inches
Ground Clearance:	11.2 inches
Approach Angle:	37 degrees
Departure Angle:	35 degrees
Weight, Combat Loaded:	20,000 pounds
Power to Weight Ratio:	14.7 hp/ton

ARMOR

Type: Rolled face hardened steel; Bolted assembly

Thickness:	Actual	Angle w/Vertical
Howitzer Shield	0.25 inches (6.4mm)	
Front, Radiator louvers	0.25 inches (6.4mm)	26 degrees
Windshield cover	0.50 inches (12.7mm)	25 degrees
Sides	0.25 inches (6.4mm)	0 degrees
Rear	0.25 inches (6.4mm)	0 degrees
Top, Hood only	0.25 inches (6.4mm)	83 degrees

ARMAMENT

Primary: 105mm howitzer M2A1 on mount T2

Traverse: Manual	40 degrees
	(20 left, 20 right)
Elevation: Manual	+35 to -5 degrees
Firing Rate: (max)	8 rounds/minute
Loading System:	Manual
Stabilizer System:	None

Secondary:

(1) .50 caliber MG HB M2 on pedestal mount M25 modified

Provision for (1) .45 caliber SMG M1928A1

Provision for (4) .30 caliber Carbines M1 or
(4) .30 caliber Rifles M1

AMMUNITION

8 rounds 105mm
300 rounds .50 caliber
500 rounds .45 caliber

FIRE CONTROL AND VISION EQUIPMENT

Primary Weapon:	Direct	Indirect
	Elbow Telescope M16	Panoramic Telescope M12A2
		Range Quadrant M4
		Gunner's Quadrant M1

Vision Devices:

Vision slots in windshield armor and front door armor
Open top vehicle

ENGINE

Make and Model: White 160AX	
Type: 6 cylinder, 4 cycle, in-line	
Cooling System: Liquid	Ignition: Battery
Displacement:	386 cubic inches
Bore and Stroke:	4 x 5.125 inches
Compression Ratio:	6.44:1
Net Horsepower: (max)	147 hp at 3000 rpm
Net Torque: (max)	325 ft-lbs at 1200 rpm
Weight:	1015 pounds, dry
Fuel: 72 octane gasoline	60 gallons
Engine Oil:	12 quarts

POWER TRAIN

Master Clutch: Dry, single plate

Transmission: Constant mesh

Gear Ratios:	1st	4.92:1	4th	1.00:1
	2nd	2.60:1	reverse	4.37:1
	3rd	1.74:1		

Transfer Case: Constant mesh, direct and underdrive

Gear Ratios:	high	1.00:1	low	2.48:1

Differential: Front axle

Gear Ratio: 6.8:1

Differential: Track drive

Gear Ratio: 4.44:1

Drive Sprocket: At front of track with 18 teeth

Pitch Diameter: 22.918 inches

RUNNING GEAR

Suspension:

Front: Semi-elliptic longitudinal leaf spring
2 ventilated disc steel wheels (1/side)
Combat Tires: 12 ply
Tire Size: 8.25 x 20

Rear: Vertical volute spring
8 dual wheels in two bogies (1 bogie/track)
Tire Size: 12 x 3.75 inches (or 12 x 4.125)
2 dual track return rollers (1/track)
Adjustable fixed idler at rear of each track
Idler Size: 18.5 x 9.375 inches
Track Type: Endless band
Track Width: 12 inches
Track Pitch: 4 inches, 58 pitches/track
Track Ground Contact Length: 46.75 inches

ELECTRICAL SYSTEM

Nominal Voltage: 12 volts DC

Main Generator: (1) 12 volts, 55 amperes, driven by main engine

Auxiliary Generator: None

Battery: (1) 12 volts

COMMUNICATIONS

Flag Set M238

FIRE PROTECTION

(1) 1 quart carbon tetrachloride

PERFORMANCE

Maximum Speed: Level road		45 miles/hour
Maximum Grade:		60 per cent
Maximum Vertical Wall:		12 inches
Maximum Fording Depth:		32 inches
Minimum Turning Circle: (diameter)		59 feet
Cruising Range: Roads	approx.	200 miles

MULTIPLE GUN MOTOR CARRIAGE M15
COMBINATION GUN MOTOR CARRIAGE M15A1
(Based upon the half track personnel carrier M3)

GENERAL DATA
Crew: 7 men
Length: w/roller   236.5 inches
Width: 98 inches
Height: 104 inches
Tread: Front   64.5 inches
       Rear   63.8 inches
Wheelbase: 135.5 inches
Ground Clearance: 11.2 inches
Approach Angle: w/roller   37 degrees
Departure Angle: 35 degrees
Weight, Combat Loaded: 20,000 pounds
Power to Weight Ratio: Net   14.7 hp/ton

ARMOR
Type: Rolled face hardened steel; Bolted assembly

Thickness:	Actual	Angle w/Vertical
Rotating Shield	0.25 inches (6.4mm)	0 degrees
Front, Radiator louvers	0.25 inches (6.4mm)	26 degrees
Windshield cover	0.50 inches (12.7mm)	25 degrees
Sides	0.25 inches (6.4mm)	0 degrees
Rear	0.25 inches (6.4mm)	0 degrees
Top, Hood only	0.25 inches (6.4mm)	83 degrees

ARMAMENT
Primary: (1) 37mm gun M1A2, (2) .50 caliber MG HB M2 in
         combination gun mounts M42 (M15) or M54 (M15A1)
    Traverse: Manual   360 degrees
    Elevation: Manual, M15   +85 to 0 degrees
               M15A1   +85 to -5 degrees
    Minimum elevation +20 degrees over the front for both mounts
    Firing Rate: (max) 37mm gun M1A2   120 rounds/minute
                 .50 caliber MG HB M2 450-575 rounds/minute/gun
    Loading System: Automatic
    Stabilizer System: None
Secondary:
    Provision for (4) .30 caliber Carbine M1

AMMUNITION
    200 rounds 37mm
    1200 rounds .50 caliber

FIRE CONTROL AND VISION EQUIPMENT
Primary Weapon: Sighting system M6 (M15)
                Computing sight M14 (M15A1)
Vision Devices:
    Vision slots in windshield armor and front door armor
    Open top vehicle

ENGINE
Make and Model: White 160AX
Type: 6 cylinder, 4 cycle, in-line
Cooling System: Liquid   Ignition: Battery
Displacement: 386 cubic inches
Bore and Stroke: 4 x 5.125 inches
Compression Ratio: 6.44:1
Net Horsepower: (max)   147 hp at 3000 rpm
Net Torque: (max)   325 ft-lbs at 1200 rpm
Weight: 1015 pounds, dry
Fuel: 72 octane gasoline   60 gallons
Engine Oil: 12 quarts

POWER TRAIN
Master Clutch: Dry, single plate
Transmission: Constant mesh

Gear Ratios:	1st	4.92:1	4th	1.00:1
	2nd	2.60:1	reverse	4.37:1
	3rd	1.74:1		

Transfer Case: Constant mesh, direct and underdrive
    Gear Ratios: high 1.00:1   low   2.48:1
Differential: Front axle
    Gear Ratio: 6.8:1
Differential: Track drive
    Gear Ratio: 4.44:1
Drive Sprocket: At front of track with 18 teeth
    Pitch Diameter: 22.918 inches

RUNNING GEAR
Suspension:
    Front: Semi-elliptic longitudinal leaf spring
        2 ventilated disc steel wheels (1/side)
        Combat tires, 12 ply
        Tire Size: 8.25 x 20
    Rear: Vertical volute spring
        8 dual wheels in two bogies (1 bogie/track)
        Tire Size: 12 x 3.75 inches (or 12 x 4.125)
        2 dual track return rollers (1/track)
        Adjustable fixed idler at rear of each track (early)
        Spring loaded idler at rear of each track (late)
        Idler Size: 18.5 x 9.375 inches
        Track Type: Endless band
        Track Width: 12 inches
        Track Pitch: 4 inches, 58 pitches/track
        Track Ground Contact Length: 46.75 inches

ELECTRICAL SYSTEM
Nominal Voltage: 12 volts DC
Main Generator: (1) 12 volts, 55 amperes, driven by main engine
Auxiliary Generator: None
Battery: (1) 12 volts

COMMUNICATIONS
Radio: SCR 593 or SCR 510

FIRE AND GAS PROTECTION
    (1) 2 pound carbon dioxide, portable
    (1) 1½ quart decontaminating apparatus M2

PERFORMANCE
Maximum Speed: Level road   45 miles/hour
Maximum Grade: 60 per cent
Maximum Vertical Wall: 12 inches
Maximum Fording Depth: 32 inches
Minimum Turning Circle: (diameter)   59 feet
Cruising Range: Roads   approx.   200 miles

GENERAL DATA

Crew:	5 men
Length: w/winch	256 inches
Width:	77.9 inches
Height: M13	88 inches
M16	103 inches
M16A1	103 inches
Tread: Front	64.5 inches
Rear	63.8 inches
Wheelbase:	135.5 inches
Ground Clearance:	11.2 inches
Approach Angle: w/winch	32 degrees
Departure Angle:	35 degrees
Weight, Combat Loaded: M13	18,500 pounds
M16	19,000 pounds
M16A1	20,000 pounds
Power to Weight Ratio: Net, M13	15.9 hp/ton
M16	15.5 hp/ton
M16A1	14.7 hp/ton
Winch Capacity:	10,000 pounds

ARMOR

Type: Rolled face hardened steel; Bolted assembly

Thickness:	Actual	Angle w/Vertical
Shields	0.25 inches (6.4mm)	0 to 35 degrees
Front, Radiator louvers	0.25 inches (6.4mm)	26 degrees
Windshield cover	0.50 inches (12.7mm)	25 degrees
Sides	0.25 inches (6.4mm)	0 degrees
Rear	0.25 inches (6.4mm)	0 degrees
Top, Hood only	0.25 inches (6.4mm)	83 degrees

ARMAMENT

Primary:

M13, (2) .50 caliber MG HB M2 (TT) on Mount M33

M16, (4) .50 caliber MG HB M2 (TT) on Mount M45D

M16A1 (4) .50 caliber MG HB M2 (TT) on Mount M45F

Traverse: Electric	360 degrees
Traverse Rate: (max)	60 degrees/second
Elevation: Electric	+90 to -10 degrees
Firing Rate: Each gun	450-575 rounds/minute
Loading System:	Automatic
Stabilizer System:	None

Secondary:

Provision for (1) .45 caliber SMG M1928A1

Provision for (1) .30 caliber Rifle M1903

Provision for (3) .30 caliber Carbine M1

Provision for (1) Grenade Launcher M1 (for rifle)

AMMUNITION

5000 rounds .50 caliber

420 rounds .45 caliber

26 hand grenades

10 rifle grenades M9A1

FIRE CONTROL AND VISION EQUIPMENT

Primary Weapon: Reflex sight M18 or illuminated sight MK9 MOD 1

Vision Devices:

Vision slots in windshield armor and front door armor

Open top vehicle

ENGINE

Make and Model: White 160AX

Type: 6 cylinder, 4 cycle, in-line

Cooling System: Liquid	Ignition: Battery	
Displacement:		386 cubic inches
Bore and Stroke:		4 x 5.125 inches
Compression Ratio:		6.44:1
Net Horsepower: (max)		147 hp at 3000 rpm
Net Torque: (max)		325 ft-lbs at 1200 rpm
Weight:		1015 pounds, dry
Fuel: 72 octane gasoline		60 gallons
Engine Oil:		12 quarts

POWER TRAIN

Master Clutch: Dry, single plate

Transmission: Constant mesh

Gear Ratios:	1st	4.92:1	4th	1.00:1
	2nd	2.60:1	reverse	4.37:1
	3rd	1.74:1		

Transfer Case: Constant mesh, direct and underdrive

Gear Ratios: high 1.00:1 low 2.48:1

Differential: Front axle

Gear Ratio: 6.8:1

Differential: Track drive

Gear Ratio: 4.44:1

Drive Sprocket: At front of track with 18 teeth

Pitch Diameter: 22.918 inches

RUNNING GEAR

Suspension:

Front: Semi-elliptic longitudinal leaf spring

2 ventilated disc steel wheels (1/side)

Combat tires, 12 ply

Tire Size: 8.25 x 20

Rear: Vertical volute spring

8 dual wheels in two bogies (1 bogie/track)

Tire Size: 12 x 3.75 inches (or 12 x 4.125)

2 dual track return rollers (1/track)

Adjustable fixed idler at rear of each track (early)

Spring loaded idler at rear of each track (late)

Idler Size: 18.5 x 9.375 inches

Track Type: Endless band

Track Width: 12 inches

Track Pitch: 4 inches, 58 pitches/track

Track Ground Contact Length: 46.75 inches

ELECTRICAL SYSTEM

Nominal Voltage: 12 volts DC

Main Generator: (1) 12 volts, 55 amperes, driven by main engine

Auxiliary Generator: None

Battery: (1) 12 volts

COMMUNICATIONS

Radio: SCR 510 or SCR 528

Flag Set M238

FIRE AND GAS PROTECTION

(1) 2 pound carbon dioxide, portable

(1) 1½ quart decontaminating apparatus M2

PERFORMANCE

Maximum Speed: Level road		45 mile/hour
Maximum Grade:		60 per cent
Maximum Vertical Wall:		12 inches
Maximum Fording Depth:		32 inches
Minimum Turning Circle: (diameter)		59 feet
Cruising Range: Roads	approx.	200 miles

MULTIPLE GUN MOTOR CARRIAGES M14 AND M17
(Based upon the half track personnel carrier M5)

GENERAL DATA

Crew:	5 men
Length: w/winch	255.5 inches
Width:	85.625 inches
Height: M14	90 inches
M17	90 inches
Tread: Front	66.5 inches
Rear	63.8 inches
Wheelbase:	135.5 inches
Ground Clearance:	11.2 inches
Approach Angle: w/winch	36 degrees
Departure Angle:	32 degrees
Weight, Combat Loaded: M14	19,200 pounds
M17	19,700 pounds
Power to Weight Ratio: M14	14.9 hp/ton
M17	14.5 hp/ton
Winch Capacity:	10,000 pounds

ARMOR

Type: Vehicle, rolled homogeneous steel; Welded assembly
   Mount, rolled face hardened steel; Bolted assembly

Thickness:	Actual	Angle w/Vertical
Shield on mount	0.25 inches (6.4mm)	0 to 35 degrees
Front, Radiator louvers	0.31 inches (7.9mm)	27 degrees
Windshield cover	0.625 inches (15.9mm)	23 degrees
Sides	0.31 inches (7.9mm)	0 degrees
Rear	0.31 inches (7.9mm)	0 degrees
Top, Hood only	0.31 inches (7.9mm)	83 degrees

ARMAMENT

Primary:
   M14, (2) .50 caliber MG HB M2 (TT) on Mount M33
   M17, (4) .50 caliber MG HB M2 (TT) on Mount M45D

Traverse: Electric	360 degrees
Traverse Rate: (max)	60 degrees/second
Elevation: Electric	+90 to -10 degrees
Firing Rate: Each gun	450-575 rounds/minute
Loading System:	Automatic
Stabilizer System:	None

Secondary:
   Provision for (1) .45 caliber SMG M1928A1
   Provision for (1) .30 caliber Rifle M1903
   Provision for (3) .30 caliber Carbine M1
   Provision for (1) Grenade Launcher M1 (for rifle)

AMMUNITION

5000 rounds .50 caliber
420 rounds .45 caliber
26 hand grenades
10 rifle grenades M9A1

FIRE CONTROL AND VISION EQUIPMENT

Primary Weapon: Reflex sight M18 or illuminated sight MK9 MOD 1
Vision Devices:
   Vision slots in windshield armor and front door armor
   Open top vehicle

ENGINE

Make and Model: International Harvester RED-450-B
Type: 6 cylinder, 4 cycle, in-line

Cooling System: Liquid	Ignition: Battery	
Displacement:		450.99 cubic inches
Bore and Stroke:		4.375 x 5 inches
Compression Ratio:		6.35:1
Net Horsepower: (max)		143 hp at 2700 rpm
Net Torque: (max)		348 ft-lbs at 800 rpm
Weight:		1250 pounds, dry
Fuel: 72 octane gasoline		60 gallons
Engine Oil:		10.5 quarts

POWER TRAIN

Master Clutch: Dry, single disc
Transmission: Constant mesh

Gear Ratios:	1st	4.92:1	4th	1.00:1
	2nd	2.60:1	reverse	4.37:1
	3rd	1.74:1		

Transfer Case: Constant mesh, direct and underdrive
   Gear Ratios: high   1.00:1   low   2.48:1
Differential: Front axle
   Gear Ratio: 7.16:1
Differential: Track drive
   Gear Ratio: 4.22:1
Drive Sprocket: At front of track with 18 teeth
   Pitch Diameter: 22.918 inches

RUNNING GEAR

Suspension:
   Front: Semi-elliptic longitudinal leaf spring
      2 ventilated disc steel wheels (1/side)
      Combat tires, 12 ply
      Tire Size: 9.00 x 20
   Rear: Vertical volute spring
      8 dual wheels in two bogies (1 bogie/track)
      Tire Size: 12 x 3.75 inches (or 12 x 4.125)
      2 dual track return rollers (1/track)
      Spring loaded idler at rear of each track
      Idler Size: 18.5 x 9.375 inches
      Track Type: Endless band
      Track Width: 12 inches
      Track Pitch: 4 inches, 58 pitches/track
      Track Ground Contact Length: 46.75 inches

ELECTRICAL SYSTEM

Nominal Voltage: 12 volts DC
Main Generator: (1) 12 volts, 55 amperes, driven by main engine
Auxiliary Generator: None
Battery: (1) 12 volts

COMMUNICATIONS

Radio: SCR 528 or British Number 19
Flag Set M238

FIRE AND GAS PROTECTION

   (1) 2 pound carbon dioxide, portable
   (1) 1½ quart decontaminating apparatus M2

PERFORMANCE

Maximum Speed: Level road		42 miles/hour
Maximum Grade:		60 per cent
Maximum Vertical Wall:		12 inches
Maximum Fording Depth:		32 inches
Minimum Turning Circle: (diameter)		59 feet
Cruising Range: Roads	approx.	200 miles

TWIN 20mm GUN MOTOR CARRIAGE T10E1
(Based upon the half-track personnel carrier M3)

GENERAL DATA

Crew:	5 men
Length: w/roller	236.75 inches
w/winch	256 inches
Width:	77.9 inches
Height:	96.75 inches
Tread: Front	64.5 inches
Rear	63.8 inches
Wheelbase:	135.5 inches
Ground Clearance:	11.2 inches
Approach Angle: w/roller	37 degrees
w/winch	32 degrees
Departure Angle:	35 degrees
Weight, Combat Loaded:	19,500 pounds
Weight, Unstowed	14,000 pounds
Power to Weight Ratio: Net	15.1 hp/ton
Winch Capacity:	10,000 pounds

ARMOR

Type: Rolled face hardened steel; Bolted assembly

Thickness:	Actual	Angle w/Vertical
Shields	0.25 inches (6.4mm)	0 to 35 degrees
Front, Radiator louvers	0.25 inches (6.4mm)	26 degrees
Windshield cover	0.50 inches (12.7mm)	25 degrees
Sides	0.25 inches (6.4mm)	0 degrees
Rear	0.25 inches (6.4mm)	0 degrees
Top, Hood only	0.25 inches (6.4mm)	83 degrees

ARMAMENT

Primary:

(2) 20mm Mark IV Oerlikon guns in Mount T17E1

Traverse: Electric	360 degrees
Traverse Rate: (max)	60 degrees/second
Elevation: Electric	+90 to -10 degrees
Firing Rate: Each gun	390-500 rounds/minute
Loading System:	Automatic
Stabilizer System:	None

Secondary:

Provision for (1) .45 caliber SMG M3
Provision for (1) .30 caliber Rifle M1903
Provision for (3) .30 caliber Carbine M1
Provision for (1) Grenade Launcher M1 (for rifle)

AMMUNITION

3000 rounds 20mm
420 rounds .45 caliber
26 hand grenades
10 rifle grenades M9A1

FIRE CONTROL AND VISION EQUIPMENT

Primary Weapon: Reflex sight Mark IX
Vision Devices:
Vision slots in windshield armor and front door armor
Open top vehicle

ENGINE

Make and Model: White 160AX	
Type: 6 cylinder, 4 cycle, in-line	
Cooling System: Liquid Ignition: Battery	
Displacement:	386 cubic inches
Bore and Stroke:	4 x 5.125 inches
Compression Ratio:	6.44:1
Net Horsepower: (max)	147 hp at 3000 rpm
Net Torque: (max)	325 ft-lbs at 1200 rpm
Weight:	1015 pounds, dry
Fuel: 72 octane gasoline	60 gallons
Engine Oil:	12 quarts

POWER TRAIN

Master Clutch: Dry, single plate
Transmission: Constant mesh

Gear Ratios:	1st	4.92:1	4th	1.00:1
	2nd	2.60:1	reverse	4.37:1
	3rd	1.74:1		

Transfer Case: Constant mesh, direct and underdrive

Gear Ratios: high   1.00:1   low   2.48:1

Differential: Front axle
Gear Ratio: 6.8:1
Differential: Track drive
Gear Ratio: 4.44:1
Drive Sprocket: At front of track with 18 teeth
Pitch Diameter: 22.918 inches

RUNNING GEAR

Suspension:
Front: Semi-elliptic longitudinal leaf spring
2 ventilated disc steel wheels (1/side)
Combat tires, 12 ply
Tire Size: 8.25 x 20
Rear: Vertical volute spring
8 dual wheels in two bogies (1 bogie/track)
Tire Size: 12 x 3.75 inches (or 12 x 4.125)
2 duel track return rollers (1/track)
Spring loaded idler at rear of each track
Idler Size: 18.5 x 9.375 inches
Track Type: Endless band
Track Width: 12 inches
Track Pitch: 4 inches, 58 pitches/track
Track Ground Contact Length: 46.75 inches

ELECTRICAL SYSTEM

Nominal Voltage: 12 volts DC
Main Generator: (1) 12 volts, 55 amperes, driven by main engine
Auxiliary Generator: None
Battery: (1) 12 volts

COMMUNICATIONS

Radio: SCR 510 or SCR 528
Flag Set M238

FIRE AND GAS PROTECTION

(1) 2 pound carbon dioxide, portable
(1) 1½ quart decontaminating apparatus M2

PERFORMANCE

Maximum Speed: Level road		45 miles/hour
Maximum Grade:		60 per cent
Maximum Vertical Wall:		12 inches
Maximum Fording Depth:		32 inches
Minimum Turning Circle: (diameter)		59 feet
Cruising Range: Roads	approx.	200 miles

40mm GUN MOTOR CARRIAGE T54E1
(Based upon the half-track personnel carrier M3)

GENERAL DATA

Crew:	6 men
Length : w/roller	236.5 inches
Width: Over side armor	77.25 inches
Height: Over rotating shield	89 inches
Tread: Front	64.5 inches
Rear	63.8 inches
Wheelbase:	135.5 inches
Ground Clearance:	11.2 inches
Approach Angle: w/roller	37 degrees
Departure Angle:	35 degrees
Weight, Combat Loaded:	22,000 pounds
Power to Weight Ratio: Net	13.4 hp/ton

ARMOR

Type: Rolled face hardened steel; Bolted assembly

Thickness:	Actual	Angle w/Vertical
Rotating Shield	0.19 inches(4.8mm)	0 degrees
Front, Radiator louvers	0.25 inches (6.4mm)	26 degrees
Windshield cover	0.50 inches (12.7mm)	25 degrees
Sides	0.25 inches (6.4mm)	0 degrees
Rear	0.25 inches (6.4mm)	0 degrees
Top, Hood only	0.25 inches (6.4mm)	83 degrees

ARMAMENT

Primary: (1) 40mm gun M1 in mount T5

Traverse: Manual	360 degrees
Elevation: Manual	+90 to -5 degrees
Firing Rate: (max)	20 rounds/minute
Loading System:	Automatic
Stabilizer System:	None

AMMUNITION

120 rounds 40mm

FIRE CONTROL AND VISION EQUIPMENT

Primary Weapon: Sighting system T15

Vision Devices:

Vision slots in windshield armor and front door armor

Open top vehicle

ENGINE

Make and Model: White 160AX	
Type: 6 cylinder, 4 cycle, in-line	
Cooling System: Liquid	Ignition: Battery
Displacement:	386 cubic inches
Bore and Stroke:	4 x 5.125 inches
Compression Ratio:	6.44:1
Net Horsepower: (max)	147 hp at 3000 rpm
Net Torque: (max)	325 ft-lbs at 1200 rpm
Weight:	1015 pounds, dry
Fuel: 72 octane gasoline	60 gallons
Engine Oil:	12 quarts

POWER TRAIN

Master Clutch: Dry, single plate

Transmission: Constant mesh

Gear Ratios:	1st	4.92:1	4th	1.00:1
	2nd	2.60:1	reverse	4.37:1
	3rd	1.74:1		

Transfer Case: Constant mesh, direct and underdrive

Gear Ratios:	high	1.00:1	low	2.48:1

Differential: Front axle

Gear Ratio: 6.8:1

Differential: Track drive

Gear Ratio: 4.44:1

Drive Sprocket: At front of track with 18 teeth

Pitch Diameter: 22.918 inches

RUNNING GEAR

Suspension:

Front: Semi-elliptic longitudinal leaf spring

2 ventilated disc steel wheels (1/side)

Combat tires (12 ply)

Tire Size: 8.25 x 20

Rear: Vertical volute spring

8 dual wheels in two bogies (1bogie/track)

Tire Size: 12 x 3.75 inches (or 12 x 4.125)

2 dual track return rollers (1/track)

Adjustable fixed idler at rear of each track

Idler Size: 18.5 x 9.375 inches

Track Type: Endless band

Track Width: 12 inches

Track Pitch: 4 inches, 58 pitches/track

Track Ground Contact Length: 46.75 inches

ELECTRICAL SYSTEM

Nominal Voltage: 12 volts DC

Main Generator: (1) 12 volts , 55 amperes, driven by main engine

Auxiliary Generator: None

Battery: (1) 12 volts

COMMUNICATIONS

Radio: None

FIRE PROTECTION

(1) 2 pound carbon dioxide, portable

PERFORMANCE

Maximum Speed: Level road	45 miles/hour
Maximum Grade:	60 per cent
Maximum Vertical Wall:	12 inches
Maximum Fording Depth:	32 inches
Minimum Turning Circle: (diameter)	59 feet
Cruising Range: Roads	approx. 200 miles

HALF-TRACK CAR T16

GENERAL DATA

Crew:	10	men
Length: Overall w/roller	258.25	inches
Width: Over side armor	77.25	inches
Height: Over roof	92.50	inches
Tread: Front	64.5	inches
Rear	63.8	inches
Wheelbase:	147.75	inches
Ground Clearance:	11.375	inches
Approach Angle: w/roller	37	degrees
Departure Angle:	48.5	degrees
Weight, Combat Loaded:	20,225	pounds
Weight, Unstowed:	16,625	pounds
Power to Weight Ratio: Net	14.5	hp/ton

ARMOR

Type: Rolled face hardened steel; Bolted assembly

Thickness:	Actual	Angle w/Vertical
Front, Radiator louvers	0.25 inches (6.4mm)	26 degrees
Windshield cover	0.50 inches (12.7mm)	25 degrees
Sides	0.25 inches (6.4mm)	0 degrees
Rear	0.25 inches (6.4mm)	0 degrees
Top, Hood	0.25 inches (6.4mm)	83 degrees
Roof	0.25 inches (6.4mm)	70 to 90 degrees

ARMAMENT

(1) .50 caliber MGHB M2 flexible on skate mount
(2) .30 caliber MG M1919A4 on skate mount
Provision for (1) .45 caliber SMG M1928A1

AMMUNITION

750 rounds .50 caliber
12,000 rounds .30 caliber
540 rounds .45 caliber
10 hand grenades

VISION EQUIPMENT

Vision slots in windshield armor and front door armor
Vehicle open under top

ENGINE

Make and Model: White 160AX
Type: 6 cylinder, 4 cycle, in-line
Cooling System: Liquid   Ignition: Battery

Displacement:	386 cubic inches
Bore and Stroke:	4 x 5.125 inches
Compression Ratio:	6.44:1
Net Horsepower: (max)	147 hp at 3000 rpm
Net Torque: (max)	325 ft-lbs at 1200 rpm
Weight:	1015 pounds, dry
Fuel: 72 octane gasoline	60 gallons
Engine Oil:	12 quarts

POWER TRAIN

Master Clutch: Dry, single plate
Transmission: Constant mesh

Gear Ratios:	1st	4.92:1	4th	1.00:1
	2nd	2.60:1	reverse	4.37:1
	3rd	1.74:1		

Transfer Case: Constant mesh, direct and underdrive
   Gear ratios: high   1.00:1   low   2.48:1
Differential: Front axle
   Gear Ratio: 6.8:1
Differential: Track drive
   Gear Ratio: 4.44:1
Drive Sprocket: At front of track with 18 teeth
   Pitch Diameter: 22.918 inches

RUNNING GEAR

Suspension:
   Front: Semi-elliptic longitudinal leaf spring
      2 ventilated disc steel wheels (1/side)
      Combat Tires:
      Tire Size: 8.50 x 20
   Rear: Vertical volute spring
      8 dual wheels in two bogies (1 bogie/track)
      Tire Size: 16 x 4.125 inches
      4 dual track return rollers (2/track)
      Adjustable fixed idler at rear of each track
      Idler Size: 18.5 x 9.375 inches
      Track Type: Endless band
      Track Width: 14 inches
      Track Pitch: 4 inches, 72 pitches/track
      Track Ground Contact Length: 61.5 inches

ELECTRICAL SYSTEM

Nominal Voltage: 12 volts DC
Main Generator: (1) 12 volts, 55 amperes, driven by main engine
Auxiliary Generator: None
Battery: (1) 12 volts

COMMUNICATIONS

Radio: SCR 193 or 506 and 508 and 593; SCR 284 and 508 and 593;
     SCR 193 or 506 and 508 or 528 or 608 or 610 or 628

FIRE AND GAS PROTECTION

(1) 2 pound carbon dioxide, portable
(2) 1½ quart decontaminating apparatus M2

PERFORMANCE

Maximum Speed: Level road		37 miles/hour
Maximum Grade:		60 per cent
Maximum Vertical Wall:		12 inches
Maximum Fording Depth:		32 inches
Minimum Turning Circle: (diameter)		59 feet
Cruising Range: Roads	approx.	200 miles

Armament on the basic half-track cars and personnel carriers consisted of .30 and .50 caliber machine guns. On the early half-track cars, both types of weapons were installed on skate mounts riding on a rail surrounding the rear compartment of the vehicle. The early personnel carriers were armed with a .30 caliber machine gun on a pedestal mount in the rear compartment. However, it was frequently replaced by a .50 caliber weapon. Later half-tracks carried the .50 caliber machine gun on an M49 ring mount fitted over the right side of the driving compartment. Three pintle mounts, located on each side and the rear, were provided for the .30 caliber machine gun. The M4 and M4A1 mortar carriers also utilized the skate mount for a .30 caliber machine gun. The later M21 mortar carrier was fitted with a .50 caliber weapon on a pedestal mount.

The widespread use of the half-track chassis as an expedient weapon carrier resulted in the installation of a considerable number of heavy weapons. These ranged in caliber up to the 105mm howitzer M2A1. Although the latter was considered to be too heavy for the chassis, it served successfully. Data describing the characteristics of the various weapons installed on the half-track chassis are tabulated in the following data sheets. The dimensions of the heavier weapons are defined in the sketch.

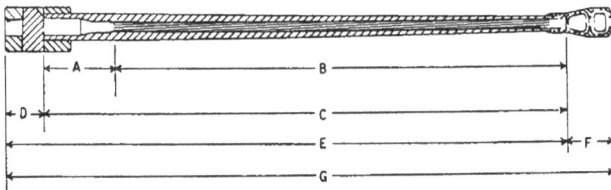

A. Length of Chamber (to rifling)
B. Length of Rifling
C. Length of Bore
D. Depth of Breech Recess
E. Length, Muzzle to Rear Face of Breech
F. Additional Length, Muzzle Brake, Etc.
G. Overall Length

The official nomenclature for each type of ammunition is that used during its period of greatest service. However, since this was frequently changed during the service life, a standard nomenclature is added in parentheses to prevent confusion. These standard terms, which are used separately and in combination are defined as follows:

AP	Armor piercing, uncapped
APBC	Armor piercing with ballistic cap
APCBC	Armor piercing with armor piercing cap and ballistic cap
HE	High explosive
HEAT	High explosive antitank, shaped charge
CP	Concrete piercing
-T	Tracer

Penetration performance data for some types of armor piercing rounds are quoted for 30 degree angles of obliquity as this was the common practice during World War II. It should be noted that the relative performance of different types of projectiles at 30 degrees is not necessarily maintained at other angles of obliquity. The simplified sketch in the introduction to the vehicle data sheets defined the angle of obliquity as the angle between a line perpendicular to the armor plate and the projectile path. However, in three dimensions, the calculation of the true angle is a little more complicated as indicated in these sketches. Here, the angle of obliquity is shown to be the angle whose cosine equals the product of the cosines for the vertical and lateral attack angles.

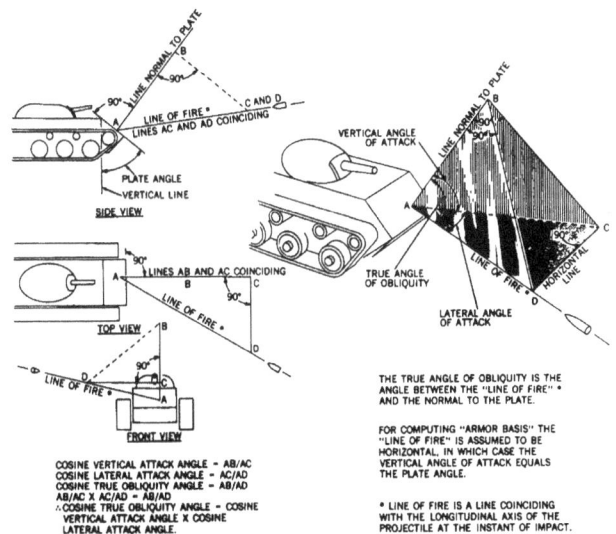

227

.30 CALIBER MACHINE GUNS M1917A1 AND M1919A4, BROWNING

Carriage and Mount	Early M2 half-track cars on skate mount (M1917A1), M2 and M3 series half-tracks on skate, pintle or pedestal mounts (M1919A4)
Length of Rifling	21.28 inches (M1917A1), 21.38 inches (M1919A4)
Length of Barrel	23.9 inches (M1917A1), 24 inches (M1919A4 flexible)
Length Overall	38.64 inches (M1917A1), 41.11 inches (M1919A4 flexible)
Diameter of Bore	.300 inches
Total Weight	32.6 pounds (M1917A1) less water, 31 pounds (M1919A4 flexible)
Weight of Recoiling Parts	7.35 pounds (M1917A1), 11.7 pounds (M1919A4)
Weight of Barrel	3 pounds (M1917A1), 7.35 pounds (M1919A4)
Operation	Short recoil
Feed System	Fabric belt, 250 rounds
Cooling System	Water (M1917A1), air (M1919A4)
Rifling	4 grooves, uniform right-hand twist, 1 turn in 33.3 calibers
Ammunition	.30 caliber M2 ball, .30 caliber M2 armor piercing (AP), .30 caliber M1 tracer, .30 caliber M1 incendiary
Powder Pressure	52,000 psi M2 ball, 54,000 psi M2 AP, 52,000 psi M1 tracer, 52,000 psi M1 incendiary
Weight, Complete Round	M2 ball 396 grains, M2 AP 414 grains, M1 tracer 396 grains, M1 incendiary 386 grains
Weight, Projectile	M2 ball 152 grains, M2 AP 162 grains, M1 tracer 152.5 grains, M1 incendiary 140 grains
Rate of Fire	450-600 rounds/minute (M1917A1), 400-550 rounds/minute (M1919A4)
Muzzle Velocity *	M2 ball 2740 feet/second, M2 AP 2715 feet/second, M1 tracer 2050 feet/second, M1 incendiary 2950 feet/second
Maximum Range	M2 ball 3500 yards, M2 AP 3160 yards, M1 tracer 3350 yards, M1 incendiary 2875 yards

* Velocity plus or minus 30 feet/second measured 78 feet from muzzle with 24 inch barrel

.50 CALIBER MACHINE GUN M2, BROWNING

Types	Air-cooled, heavy barrel (HB); Water-cooled (WC); Air-cooled, aircraft (AC)
Carriage and Mount	M2 and M3 series half-tracks (HB) motor carriages T28, T28E1 (WC); M15, M15A1, M13, M14, M16, M16A1, M16A2, M17 (HB); T1E2, T37, T37E1 (AC)
Length of Rifling	40.91 inches (HB), 40.11 inches (WC), 31.91 inches (AC)
Length of Barrel	45 inches (HB), 45 inches (WC), 36 inches (AC)
Length Overall	65 inches (HB), 66 inches (WC), 57 inches (AC)
Diameter of Bore	0.500 inches
Total Weight	82 pounds (HB fixed), 84 pounds (HB flexible), 81 pounds (HB turret) 100 pounds (WC) less water, (water 21 pounds) 64 pounds (AC fixed), 65.1 pounds (AC flexible)
Weight of Recoiling Parts	38.8 pounds (HB), 24.5 pounds (WC), 19.2 pounds (AC)
Weight of Barrel	27 pounds (HB), 15.2 pounds (WC), 9.8 pounds (AC)
Rifling	8 grooves, uniform right-hand twist, 1 turn in 30 calibers
Operation	Short recoil
Feed System	Metallic link belt, 110 round increments
Cooling System	Air (HB and AC), water (WC),
Ammunition	.50 caliber M2 ball, .50 caliber M1 tracer, .50 caliber M1 armor piercing (AP)
Weight, Complete Round	M2 ball 1830 grains, M1 tracer 1789 grains, M1 AP 1837 grains
Weight, Projectile	M2 ball 711.5 grains, M1 tracer 681 grains, M1 AP 718 grains
Powder Pressure	52,000 psi M2 ball, 52,000 psi M1 tracer, 52,000 psi M1 AP
Rate of Fire	450-575 rounds/minute (HB), 500-650 rounds/minute (WC), 750-850 rounds/minute (AC)
Muzzle Velocity *	M2 ball 2935 feet/second, M1 tracer 2865 feet/second, M1 AP 2985 feet/second
Maximum Range	M2 ball 7200 yards, M1 tracer 6000 yards, M1 AP 7200 yards

* Velocity plus or minus 30 feet/second measured 78 feet from muzzle with 36 inch barrel

20mm AUTOMATIC GUN MARK IV, OERLIKON

Carriage and Mount	Twin 20mm Gun Motor Carriages T10 and T10E1 on Mounts T17 and T17E1
Length of Rifling	48.728 inches
Length of Barrel	57 inches (approx.), 77 calibers
Length Overall	87 inches
Diameter of Bore	0.787 inches
Weight of Barrel	37 pounds
Total Weight	147 pounds
Operation	Blowback
Feed System	60 round drum type magazine
Cooling System	Air
Rifling	9 grooves, uniform right-hand twist, 1 turn in 36 calibers
Ammunition	Fixed
Weight, Complete Round	HE/T Mark IV .524 pounds
	HE Mark III .526 pounds
Weight, Projectile	HE/T Mark IV .2621 pounds
	HE Mark III .2714 pounds
Powder Pressure	48,160 psi
Rate of Fire	390-500 rounds/minute
Muzzle Velocity	2725 feet/second
Maximum Range	5000 yards (approx.)

37mm AUTOMATIC GUN M1A2

Carriage and Mount	Multiple Gun Motor Carriages T28E1, M15 and Combination Gun Motor Carriage M15A1	
Length of Chamber (to rifling)	9.85 inches	
Length of Rifling	68.35 inches	
Length of Barrel	78.2 inches	
Diameter of Bore	1.457 inches	
Chamber Capacity	17.80 cubic inches (M54 shell)	
Weight of Tube	119 pounds	
Total Weight of Gun	365 pounds	
Operation	Long recoil	
Type of Breechblock	Vertical sliding wedge	
Automatic Loader	Feed box loaded from 10 round clips	
Ammunition	Fixed	
Primer	Percussion	
Weight, Complete Round	HE-T, SD, Shell M54 (HE-T)	2.67 pounds
	APC-T Shot M59A1 (APC-T)	3.17 pounds
Weight, Projectile	HE-T, SD, Shell M54 (HE-T)	1.34 pounds
	APC-T Shot. M59A1 (APC-T)	1.91 pounds
Powder Pressure	36,000 psi	
Rate of Fire	120 rounds/minute	
Muzzle Velocity	HE-T, SD, Shell M54 (HE-T)	2600 ft/sec
	APC-T Shot M59A1 (APC-T)	2050 ft/sec
Muzzle Energy of Projectile $KE = \frac{1}{2} MV^2$ Rotational energy is neglected and values are based on long tons (2240 pounds)	HE-T, SD, Shell M54 (HE-T)	63 ft-tons
	APC-T Shot M59A1 (APC-T)	56 ft-tons
Maximum Range, Vertical*	HE-T, SD, Shell M54 (HE-T)	6300 yards
Horizontal*		9050 yards
Vertical	APC-T Shot M59A1 (APC-T)	4000 yards
Horizontal		5790 yards

Penetration Performance Homogeneous steel armor at 30 degrees obliquity

Range	500 yards	1000 yards	1500 yards	2000 yards
APC-T Shot M59A1 (APC-T)	0.9 inches (23mm)	0.7 inches (18mm)	0.6 inches (15mm)	0.5 inches (13mm)

Face-hardened steel armor at 30 degrees obliquity

Range	500 yards	1000 yards	1500 yards	2000 yards
APC-T Shot M59A1 (APC-T)	1.0 inches (25mm)	0.7 inches (18mm)	0.6 inches (15mm)	0.5 inches (13mm)

* Actual range is limited by shell destroying (SD) tracer to approximately 3500 yards

37mm GUN M3

Carriage and Mount	37mm Gun Motor Carriage M6 on Mount M25
	Improvised field installation on Half-track Car M2
Length of Chamber (to rifling)	9.55 inches
Length of Rifling	68.45 inches
Length of Chamber (to projectile base)	8.1 inch (square base projectiles)
Travel of Projectile in Bore	69.9 inches
Length of Bore	78.0 inches, 53.5 calibers
Depth of Breech Recess	4.5 inches
Length, Muzzle to Rear Face of Breech	82.5 inches, 56.6 calibers
Additional Length, Muzzle Brake, etc.	None
Overall Length	82.5 inches
Diameter of Bore	1.457 inches (37mm)
Chamber Capacity	19.35 cubic inches (APC M51), 19.19 cubic inches (HE M63)
Weight of Gun	191 pounds
Type of Breechblock	Manually operated vertical sliding wedge
Rifling	12 grooves, uniform right-hand twist, one turn in 25 calibers
Ammunition	Fixed
Primer	Percussion
Weight, Complete Round	APC M51 Shot (APCBC-T) 3.48 pounds
	AP M74 Shot (AP-T) 3.34 pounds
	HE M63 Shell (HE) 3.13 pounds
	Canister M2 3.49 pounds
Weight, Projectile	APC M51 Shot (APCBC-T) 1.92 pounds
	AP M74 Shot (AP-T) 1.92 pounds
	HE M63 Shell (HE) 1.61 pounds
	Canister M2 (122 steel balls) 1.94 pounds
Maximum Powder Pressure	50,000 psi
Maximum Rate of Fire	25 rounds/minute
Muzzle Velocity	APC M51 Shot (APCBC-T) 2900 ft/sec
	AP M74 Shot (AP-T) 2900 ft/sec
	HE M63 Shell (HE) 2600 ft/sec
	Canister M2 2500 ft/sec
Muzzle Energy of Projectile, KE=$\frac{1}{2}$MV^2	APC M51 Shot (APCBC-T) 112 ft-tons
Rotational energy is neglected and	AP M74 Shot (AP-T) 112 ft-tons
values are based on long tons	HE M63 Shell (HE) 75 ft-tons
(2240 pounds)	Canister M2 84 ft-tons
Maximum Range (independent of mount)	APC M51 Shot (APCBC-T) 12,850 yards
	AP M74 Shot (AP-T) 8,725 yards
	HE M63 Shell (HE) 9,500 yards
	Canister M2 approx. 150 to 200 yards

Penetration Performance — Homogeneous steel armor at 30 degrees obliquity

Range	500 yards	1000 yards	1500 yards	2000 yards
APC M51 Shot (APCBC-T)	2.1 inches (53mm)	1.8 inches (46mm)	1.6 inches (40mm)	1.4 inches (35mm)

Face-hardened steel armor at 30 degrees obliquity

Range	500 yards	1000 yards	1500 yards	2000 yards
APC M51 Shot (APCBC-T)	1.8 inches (46mm)	1.6 inches (40mm)	1.5 inches (38mm)	1.3 inches (33mm)

Carriage and Mount	40mm Gun Motor Carriages T1, T54, T54E, T59E1, Multiple Gun Motor Carriage T60, Multiple Gun Motor Carriage T60E1, and Twin 40mm Gun Motor Carriage T68
Length of Chamber (to rifling)	12.73 inches
Length of Rifling:	75.85 inches
Length of Chamber (to projectile base)	11.2 inches (square base shot AP-T M81A1)
	9.8 inches (boat-tailed shell HE-T Mk II)
Travel of Projectile in Bore	77.4 inches (square base shot AP-T M81A1)
	78.8 inches (boat-tailed shell HE-T Mk II)
Length of Bore	88.58 inches, 56.3 calibers
Depth of Breech Recess	5.9 inches, approx.
Length, Muzzle to Rear Face of Breech	95 inches, approx.
Additional Length, Flash Hider	10 inches, approx.
Length of Automatic Loader Assembly	39 inches, approx.
Overall Length	144 inches, approx.
Diameter of Bore	1.573 inches (40mm)
Chamber Capacity	29.9 cubic inches
Weight of Barrel Assembly	295.85 inches
Weight of Tipping Parts	1051 pounds
Type of Breechblock	Semiautomatic, vertical sliding wedge
Rifling	16 grooves, increasing twist, one turn in 45 to 30 calibers
Automatic Loader	7 round magazine loaded from 4 round clips
Ammunition	Fixed
Weight, Complete Round	AP-T M81A1 Shot (AP-T) 4.57 pounds
	HE-T,SD, Mk II Shell (HE-T) 4.70 pounds
Weight, Projectile	AP-T M81A1 Shot (AP-T) 1.96 pounds
	HE-T,SD, Mk II Shell (HE-T) 1.93 pounds
Ammunition	Fixed
Primer	Percussion
Maximum Rate of Fire	120 rounds/minute
Muzzle Velocity	AP-T M81A1 Shot (AP-T) 2870 ft/sec
	HE-T,SD, Mk II Shell (HE-T) 2870 ft/sec
Muzzle Energy of Projectile KE=$\frac{1}{2}$MV^2	AP-T M81A1 Shot (AP-T) 112 ft-tons
Rotational energy is neglected and values are based on long tons (2240 pounds)	HE-T,SD, Mk II Shell (HE-T) 110 ft-tons
Maximum Range (independent of mount)	AP-T M81A1 Shot (AP-T) 9,475 yard
	HE-T,SD, Mk II Shell (HE-T) 10,850 yards *

Penetration Performance   Homogeneous steel armor at 30 degrees obliquity

Range	500 yards	1000 yards	1500 yards	2000 yards
AP-T M81A1 Shot (APT-T)	1.9 inches (48mm)	1.6 inches (41mm)	1.2 inches (30mm)	1.0 inches (25mm)

Face-hardened steel armor at 30 degrees obliquity

Range	500 yards	1000 yards	1500 yards	2000 yards
AP-T M81A1 Shot (AP-T)	1.8 inches (46mm)	1.5 inches (38mm)	1.2 inches (30mm)	1.0 inches (25mm)

*Actual range is limited by shell destroying (SD) tracer to approximately 5200 yards horizontal and 5100 yards vertical

57mm GUN M1 AND MARK III 6 POUNDER (British)

Carriage and Mount	57mm Gun Motor Carriage T48 (M1), Pilot T48 (Mark III), in Mount T5
Length of Chamber (to rifling)	18.02 inches
Length of Rifling	94.18 inches (M1), 78.18 inches (Mark III)
Length of Chamber (to projectile base)	16.2 inches
Travel of Projectile in Bore	96.0 inches (M1), 80.0 inches (Mark III)
Length of Bore	112.20 inches, 50.0 calibers (M1); 96.20 inches, 42.9 calibers (Mark III)
Depth of Breech Recess	4.75 inches
Length, Muzzle to Rear Face of Breech	116.95 inches, 52.1 calibers (M1); 100.95 inches, 45.0 calibers (Mark III)
Additional Length, Muzzle Brake, etc.	None
Overall Length	116.95 inches (M1), 100.95 inches (Mark III)
Diameter of Bore	2.244 inches (57mm)
Chamber Capacity	100.05 cubic inches (AP-T M70), 98.87 cubic inches (APC-T M86)
Total Weight	727.5 pounds (M1), 761 pounds (Mark III)
Type of Breechblock	Semiautomatic vertical sliding wedge
Rifling	24 grooves, uniform right-hand twist, one turn in 30 calibers
Ammunition	Fixed

Weight, Complete Round		
AP-T M70 Shot (AP-T)	12.92 pounds	
APC-T M86 Projectile (APCBC/HE-T)	13.88 pounds	
Weight, Projectile		
AP-T M70 Shot (AP-T)	6.28 pounds	
APC-T M86 Projectile (APCBC/HE-T)	7.27 pounds	

Maximum Powder Pressure	46,000psi
Maximum Rate of Fire	30 rounds/minute

Muzzle Velocity	M1 gun	Mark III gun
AP-T M70 Shot (AP-T)	2950 ft/sec	2830 ft/sec
APC-T M86 Projectile (APCBC/HE-T)	2700 ft/sec	2580 ft/sec

Muzzle Energy of Projectile, KE=$\frac{1}{2}MV^2$ Rotational energy is neglected and values are based on long tons (2240 pounds)	M1 gun	Mark III gun
AP-T M70 Shot (AP-T)	349 ft-tons	379 ft-tons
APC-T M86 Projectile (APCBC/HE-T)	367 ft-tons	335 ft-tons

Maximum Range (independent of mount)		
AP-T M70 Shot (AP-T)	9,300 yards	8,900 yards
APC-T M86 Projectile (APCBC/HE-T)	13,500 yards	13,000 yards

Penetration Performance, M1 gun

Homogeneous steel armor at 30 degrees obliquity

Range	500 yards	1000 yards	1500 yards	2000 yards
APC-T M86 Projectile (APCBC/HE-T)	3.2 inches (81mm)	2.9 inches (74mm)	2.5 inches (63mm)	2.2 inches (56mm)

Face-hardened steel armor at 30 degrees obliquity

Range	500 yards	1000 yards	1500 yards	2000 yards
APC-T M86 Projectile (APCBC/HE-T)	3.0 inches (76mm)	2.9 inches (74mm)	2.7 inches (68mm)	2.5 inches (63mm)

Carriage and Mount	75mm Howitzer Motor Carriage T30 in Mount T10
Length of Chamber (to rifling)	11.2 inches
Length of Rifling	35.91 inches
Length of Chamber (to projectile base)	7.7 inches (boat-tailed projectiles)
Travel of Projectile in Bore	39.3 inches (boat-tailed projectiles)
Length of Bore	47.03 inches, 15.9 calibers
Depth of Breech Recess	7.15 inches
Length, Muzzle to Rear Face of Breech	54.18 inches, 18.4 calibers
Additional Length, Muzzle Brake, etc.	None
Overall Length	54.18 inches
Diameter of Bore	2.950 inches (75mm)
Chamber Capacity	59.08 cubic inches (HE M48)
Weight of Tube	221 pounds
Weight of Howitzer w/Breech Mechanism	341 pounds
Type of Breechblock	Manually operated horizontal sliding wedge
Rifling	28 grooves, uniform right-hand twist, one turn in 20 calibers
Ammunition	Semifixed and Fixed (HEAT M66)
Primer	Percussion

Weight, Complete Round		
	HE M48 Shell (HE)	18.24 pounds
	HE M41A1 Shell (HE)	17.40 pounds
	HEAT M66 Shell (HEAT-T)	16.30 pounds
	WP M64 Shell, Smoke	18.89 pounds

Weight, Projectile		
	HE M48 Shell (HE)	14.60 pounds
	HE M41A1 Shell (HE)	13.76 pounds
	HEAT M66 Shell (HEAT-T)	13.10 pounds
	WP M64 Shell, Smoke	15.25 pounds

Maximum Powder Pressure	26,000 psi
Maximum Rate of Fire	6 rounds/minute for first four minutes

Muzzle Velocity		
	HE M48 Shell (HE), Charge 4	1250 ft/sec
	HE M41A1 Shell (HE), Charge 4	1250 ft/sec
	HEAT M66 Shell (HEAT-T)	1000 ft/sec
	WP M64 Shell, Smoke, Charge 4	1250 ft/sec

Muzzle Energy of Projectile, KE=$\frac{1}{2}MV^2$ Rotational energy is neglected and values are based on long tons (2240 pounds)		
	HE M48 Shell (HE), Charge 4	158 ft-tons
	HE M41A1 Shell (HE), Charge 4	149 ft-tons
	HEAT M66 Shell (HEAT-T)	91 ft-tons
	WP M64 Shell, Smoke, Charge 4	165 ft-tons

Maximum Range (independent of mount)		
	HE M48 Shell (HE), Charge 4	9620 yards
	HE M41A1 Shell (HE), Charge 4	9650 yards
	HEAT M66 Shell (HEAT-T)	7900 yards
	WP M64 Shell, Smoke, Charge 4	9620 yards

Penetration Performance	Homogeneous steel armor at 0 degrees obliquity
HEAT M66 Shell (HEAT-T)	3.6 inches at any range

75mm GUN M1897A4

Carriage and Mount	75mm Gun Motor Carriage M3 in Mount M3
	75mm Gun Motor Carriage M3A1 in Mount M5
Length of Chamber (to rifling)	14.4 inches
Length of Rifling	87.37 inches
Length of Chamber (to projectile base)	12.78 inches (APC M61)
Travel of Projectile in Bore	88.99 inches (APC M61)
Length of Bore	101.77 inches, 34.5 calibers
Overall Length	107.13 inches
Diameter of Bore	2.950 inches
Chamber Capacity	88.05 cubic inches (APC M61), 80.57 cubic inches (HE M48)
Total Weight	1026 pounds
Type of Breechblock	Nordenfeld
Rifling	24 grooves, uniform right-hand twist, one turn in 25.59 calibers (slope 7 degrees)
Ammunition	Fixed
Primer	Percussion

Weight, Complete Round	APC M61 Projectile (APCBC/HE-T)	19.92 pounds
	AP M72 Shot (AP-T)	18.80 pounds
	HE M48 Shell (HE), Supercharge	19.56 pounds
Weight, Projectile	APC M61 Projectile (APCBC/HE-T)	14.96 pounds
	AP M72 Shot (AP-T)	13.94 pounds
	HE M48 Shell (HE)	14.70 pounds
Maximum Powder Pressure	38,000psi	
Maximum Rate of Fire	6 rounds/minute	
Muzzle Velocity	APC M61 Projectile (APCBC/HE-T)	2000 ft/sec
	AP M72 Shot (AP-T)	2000 ft/sec
	HE M48 Shell (HE), Supercharge	1950 ft/sec
Muzzle Energy of Projectile, $KE=\frac{1}{2}MV^2$	APC M61 Projectile (APCBC/HE-T)	415 ft-tons
Rotational energy is neglected and	AP M72 Shot (AP-T)	387 ft-tons
values are based on long tons	HE M48 Shell (HE)	387 ft-tons
(2240 pounds)		
Maximum Range (independent of mount)	APC M61 Projectile (APCBC/HE-T)	13,870 yards
	AP M72 Shot (AP-T)	10,520 yards
	HE M48 Shell (HE), Supercharge	13,870 yards

Penetration Performance — Homogeneous steel armor at 30 degrees obliquity

Range	500 yards	1000 yards	1500 yards	2000 yards
APC M61 Projectile (APCBC/HE-T)	2.5 inches (64mm)	2.3 inches (58mm)	2.1 inches (53mm)	1.9 inches (48mm)
AP M72 Shot (AP-T)	2.9 inches (74mm)	2.4 inches (61mm)	1.9 inches (48mm)	1.5 inches (38mm)

Face-hardened steel armor at 30 degrees obliquity

Range	500 yards	1000 yards	1500 yards	2000 yards
APC M61 Projectile (APCBC/HE-T)	2.8 inches (71mm)	2.5 inches (64mm)	2.3 inches (58mm)	2.0 inches (51mm)
AP M72 Shot (AP-T)	2.5 inches (64mm)	2.0 inches (51mm)	1.5 inches (38mm)	1.2 inches (30mm)

Carriage and Mount	Proposed for installation in 105mm Howitzer Motor Carriage T38
Length of Chamber (to rifling)	15.0 inches
Length of Rifling	51.0 inches
Length of Chamber (to projectile base)	11.4 inches (boat-tailed projectiles)
Travel of Projectile in Bore	54.6 inches (boat-tailed projectiles)
Length of Bore	66.0 inches, 16.0 calibers
Depth of Breech Recess	8.3 inches
Length, Muzzle to Rear Face of Breech	74.3 inches, 17.9 calibers
Additional Length, Muzzle Brake, etc.	None
Overall Length	74.3 inches
Diameter of Bore	4.134 inches (105mm)
Chamber Capacity	153.80 cubic inches
Total Weight	955 pounds
Type of Breechblock	Manually operated horizontal sliding wedge
Rifling	36 grooves, uniform right-hand twist, one turn in 20 calibers
Ammunition	Semfixed variable charge or fixed (HEAT M67)
Primer	Percussion

Weight, Complete Round	HE M1 Shell (HE), Charge 5	40.46 pounds
	HEAT M67 Shell (HEAT-T)	36.65 pounds
	HC BE M84 Shell, Smoke, Charge 5	40.32 pounds
	WP M60 Shell, Smoke, Charge 5	41.83 pounds
Weight, Projectile	HE M1 Shell (HE)	33.00 pounds
	HEAT M67 Shell (HEAT-T)	29.22 pounds
	HC BE M84 Shell, Smoke	32.87 pounds
	WP M60 Shell, Smoke	34.31 pounds
Maximum Powder Pressure	28,000 psi	
Maximum Rate of Fire	8 rounds/minute	
Muzzle Velocity	HE M1 Shell (HE), Charge 5	1020 ft/sec
	HEAT M67 Shell (HEAT-T)	1020 ft/sec
	HC BE M84 Shell, Smoke, Charge 5	1020 ft/sec
	WP M60 Shell, Smoke, Charge 5	1020 ft/sec
Muzzle Energy of Projectile, KE=$\frac{1}{2}MV^2$	HE M1 Shell (HE), Charge 5	238 ft-tons
Rotational energy is neglected and	HEAT M67 Shell (HEAT-T)	211 ft-tons
values are based on long tons	HC BE M84 Shell, Smoke, Charge 5	237 ft-tons
(2240 pounds)	WP M60 Shell, Smoke, Charge 5	247 ft-tons
Maximum Range (independent of mount)	HE M1 Shell (HE), Charge 5	8295 yards
	HEAT M67 Shell (HEAT-T)	8490 yards
	HC BE M84 Shell, Smoke, Charge 5	8295 yards
	WP M60 Shell, Smoke, Charge 5	8295 yards

Penetration Performance
    HEAT M67 Shell (HEAT-T)

Homogeneous steel armor at 0 degrees obliquity
    4.0 inches at any range
Concrete at 0 degrees obliquity

Range	0 yards	500 yards	1000 yards	2000 yards
HE M1 Shell (HE), Charge 5 w/Concrete Piercing Fuze M78 (T105)	1.0 feet	0.9 feet	0.8 feet	0.7 feet

235

105mm HOWITZER M2A1

Carriage and Mount	105mm Howitzer Motor Carriage T19 in Mount T2
Length of Chamber (to rifling)	15.03 inches
Length of Rifling	78.02 inches
Length of Chamber (to projectile base)	11.38 inches (boat-tailed projectiles)
Travel of Projectile in Bore	81.67 inches (boat-tailed projectiles)
Length of Bore	93.05 inches, 22.5 calibers
Depth of Breech Recess	8.30 inches
Length, Muzzle to Rear Face of Breech	101.35 inches, 24.5 calibers
Additional Length, Muzzle Brake, etc.	None
Overall Length	101.35 inches
Diameter of Bore	4.134 inches (105mm)
Chamber Capacity	153.80 cubic inches
Total Weight	1080 pounds
Type of Breechblock	Manually operated horizontal sliding wedge
Rifling	36 grooves, uniform right-hand twist, one turn in 20 calibers
Ammunition	Semifixed variable charge or fixed (HEAT M67)
Primer	Percussion

Weight, Complete Round	HE M1 Shell (HE), Charge 7	42.07 pounds
	HEAT M67 Shell (HEAT-T)	36.85 pounds
	HC BE M84 Shell, Smoke, Charge 7	41.94 pounds
	WP M60 Shell, Smoke, Charge 7	43.77 pounds
Weight, Projectile	HE M1 Shell (HE)	33.00 pounds
	HEAT M67 Shell (HEAT-T)	29.22 pounds
	HC BE M84 Shell, Smoke	32.87 pounds
	WP M60 Shell, Smoke	34.31 pounds
Maximum Powder Pressure	28,000 psi	
Maximum Rate of Fire	8 rounds/minute	
Muzzle Velocity	HE M1 Shell (HE), Charge 7	1550 ft/sec
	HEAT M67 Shell (HEAT-T)	1250 ft/sec
	HC BE M84 Shell, Smoke, Charge 7	1550 ft/sec
	WP M60 Shell, Smoke, Charge 7	1550 ft/sec
Muzzle Energy of Projectile, $KE = \frac{1}{2}MV^2$	HE M1 Shell (HE), Charge 7	550 ft-tons
Rotational energy is neglected and	HEAT M67 Shell (HEAT-T)	317 ft-tons
values are based on long tons	HC BE M84 Shell, Smoke, Charge 7	547 ft-tons
(2240 pounds)	WP M60 Shell, Smoke, Charge 7	571 ft-tons
Maximum Range (independent of mount)	HE M1 Shell (HE), Charge 7	12,205 yards
	HEAT M67 Shell (HEAT-T)	8,590 yards
	HC BE M84 Shell, Smoke, Charge 7	12,205 yards
	WP M60 Shell, Smoke, Charge 7	12,150 yards

Penetration Performance
   HEAT M67 Shell (HEAT-T)
   Homogeneous steel armor at 0 degrees obliquity
      4.0 inches at any range
   Concrete at 0 degrees obliquity

Range	0 yards	500 yards	1000 yards	2000 yards
HE M1 Shell (HE), Charge 7 w/Concrete Piercing Fuze M78 (T105)	1.5 feet	1.4 feet	1.3 feet	1.1 feet

81mm MORTAR M1

Carriage and Mount	81mm Mortar Carriers M4, M4A1, and M21 w/Mount M1 for ground use
Length Overall of Barrel	49.5 inches
Diameter of Bore	3.2 inches
Rifling	None, smooth bore
Weight of Mortar	44.5 pounds
Weight w/ Mount M1	136.0 pounds, for ground use
Rate of Fire, Normal	18 rounds/minute
Maximum	30-35 rounds/minute
Ammunition	Semi-fixed, adjustable charge
Weight, Complete Round	M43A1 Shell HE	6.87 pounds
	M36 Shell HE	10.62 pounds,
	M57 WP Shell smoke	10.75 pounds
	M301 Shell illuminating	10.71 pounds
Range	M43A1 Shell HE	100 to 3290 yards
	M36 Shell HE	300 to 2558 yards
	M57 WP Shell smoke	300 to 2470 yards
	M301 Shell illuminating	2300 yards max

4.2 inch MORTAR M2

Carriage and Mount	4.2 inch Mortar Carriers T21 and T21E1
Length of Rifling	42 inches
Length Overall of Barrel	48 inches
Diameter of Bore	4.2 inches
Weight of Mortar Barrel	100 pounds
Rifling	24 grooves, increasing twist, 1 turn in 0 to 20 calibers
Ammunition	Semi-fixed, adjustable charge
Weight, Complete Round	WP M2 Shell smoke	25.5 pounds
	HE M3 Shell	24.5 pounds
Powder Pressure	13,500 psi (max. charge)
Rate of Fire, Normal	5 rounds/minute
	20 rounds/minute
Muzzle Velocity (max)	WP M2 Shell smoke	820 ft/sec
	HE M3 Shell	841 ft/sec
Range (max)	WP M2 Shell smoke	4300 yards
	HE M3 Shell	4400 yards

REFERENCES AND SELECTED BIBLIOGRAPHY

Books and Manuscripts

Gill, Lonnie, "Tank Destroyer Forces - WWII", Turner Publishing Company, Paducah, Kentucky, 1992

Houston, Donald E., "Hell on Wheels, The 2nd Armored Division", Presidio Press, San Rafael, California, 1977

Larson, Lieutenant Colonel Werner L. and Fisher, Lieutenant Colonel John R., "It was Patton's Idea, The History of an Unusual World War II Unit", Published by the 443rd AAA Association, 1984

Haugh, David R., "U.S. Half-Tracks, Their Design and Development", Darlington Productions, Darlington, Maryland, 1999

Howe, George F., "The Battle History of the 1st Armored Division", Combat Forces Press, Washington, D.C., 1954

_____ , "The Development Record in Combat Vehicles" History of the Ordnance Corps Chapter VI, manuscript 1947-48

Reports and Official Documents

"Comparative Engineering Tests of Half-Track Personnel Carriers M3 and M5", General Motors Proving Ground, Milford, Michigan, 30 March 1944

"Design, Development, Engineering, and Production of Half-Track Vehicles, 1940-1944", Office Chief of Ordnance-Detroit, Army Services Forces

"Engineering and Endurance Test of Half-Track Cars M2E5", General Motors Proving Ground, Milford, Michigan, 29 January 1943

"Field Manual FM17-71 Crew Drill for Half-Track Vehicles Armored Infantry", War Department, Washington, D.C., 31 January 1945

"Field Manual FM23-60 Browning Machine Gun, Caliber .50, HB M2 Ground" War Department, Washington, D.C., 25 September 1940

"Field Manual FM44-2 Antiaircraft Artillery Automatic Weapons", Department of the Army, Washington, D.C., 24 August 1950

"Field Manual FM44-57 Service of the Piece Multiple Machine Gun Mounts" War Department, Washington, D.C., 29 January 1945

"Final Report on Test of Half-Track Car T16 (Autocar)", 8 September 1942

"First Partial Report on the Half-Track Personnel Carrier T7", Aberdeen Proving Ground, Maryland, 1 December 1938

"First Partial Report on the Half-Track Trucks T9 and T9E1 (Production)", Aberdeen Proving Ground, Maryland, 4 April 1938

"First Partial Report on Test of Half-Track Truck T17" The Armored Board, Fort Knox, Kentucky, 4 November 1943

"First Report on Test of Half-Track Scout Car T14 (White Motor Company Pilot)", Aberdeen Proving Ground, Maryland, 13 November 1940

"First Report on Half-Track Car M2 (White Motor Company Production Pilot)", Aberdeen Proving Ground, Maryland, 3 October 1941

"First Report of Half-Track Personnel Carrier M3 No. 1904 (Diamond T Company)", Aberdeen Proving Ground, Maryland, 13 October 1941

"First Report of Three-quarter Track Trucks T16 and T19" Aberdeen Proving Ground, Maryland, 27 June 1944

"Half-Track Trucks T16, T17, and T19", Office Chief of Ordnance-Detroit, 31 December 1945

"Record of Development of Armored Cars, Half-Tracks, Tank Transporters, Tractors and Motor Transport from 1 September 1942 to 1 September 1944", 28 September 1944

"Record of the Development of Armored Cars, Half-Tracks, Tank Transporters, Trucks and Trailers from 1 September 1944 to 15 May 1945 and Record of Development of Motorcycles and Bicycles from 1 September 1942 to 15 May 1945", 28 May 1945

"Standard Nomenclature List SNL G-102 Service Parts Catalog for Cars, Half-Track M2 and M2A1; Carriers Personnel, Half-Track, M3 and M3A1; Carriages, Motor, 75mm Gun M3 and M3A1; Carriers, 81mm Mortar, Half-Track, M4 and M4A1; Carrier 81mm Mortar, Half-Track,

M21; Carriage, Motor, Multiple Gun, M13; Carriages, Motor, Multiple Gun M15 and M15A1; Carriage, Motor, Multiple Gun, M16; Carriage, Motor, 105mm Howitzer T19; Carriage, Motor, 75mm Howitzer T30; Carriage, Motor, 57mm Gun T48", War Department, Washington, D.C., 30 September 1943

"Technical Manual TM3-320 Mortar, Chemical, 4.2 inch", War Department, Washington, D.C., 30 June 1945

"Technical Manual TM9-252 40mm Automatic Gun M1 (AA) and 40mm Antiaircraft Gun Carriages M2 and M2A1", War Department, Washington, D.C., 17 January 1944

"Technical Manual TM9-303 57mm Gun M1 and Gun Carriages M1, M1A1, and M1A2", War Department, Washington, D.C., 24 February 1943

"Technical Manual TM9-303 57mm Guns M1 and Mk. III (British) and Carriages M1, M1A1, M1A2, M1A3, and M2", War Department, Washington, D.C., 25 April 1944

"Technical Manual TM9-318 75mm Howitzers M2 and M3 (Mounted in Combat Vehicles)", War Department, Washington, D.C. 14 December 1944

"Technical Manual TM9-319 75mm Pack Howitzer M1A1 and Carriage M8", Department of the Army, Washington, D.C., 17 November 1948

"Technical Manual TM9-325 105mm Howitzer M2A1, Carriages M2A1 and M2A2, and Combat Vehicle Mounts M4 and M4A1", Department of the Army, Washington, D.C., 7 May 1948

"Technical Manual TM9-707 Basic Half-Track Vehicles (IHC)", War Department, Washington, D.C., 1943

"Technical Manual TM9-710 Basic Half-Track Vehicles (White, Autocar, and Diamond T)", War Department, Washington, D.C., February 1944

"Technical Manual TM9-710 Half-Track Vehicles, Car M2A1; Personnel Carriers M3 and M3A1 81mm Mortar Carriers M4, M4A1, and M21; Combination Gun Motor Carriage M15A1 and Multiple Gun Motor Carriages M16 and M16A1", Department of the Army, Washington, D.C. 8 May 1953

"Technical Manual TM9-710C Chassis and Body for Half-Track Vehicles", War Department, Washington, D.C., 11 September 1942

"Technical Manual TM9-1707A Ordnance Maintenance Engine, Engine Accessories, and Electrical Systems for Basic Half-Track Vehicles (IHC)", War Department, Washington, D.C., 1943

"Technical Manual TM9-1707B Ordnance Maintenance Power Train, Body, and Chassis for Basic Half-Track Vehicle (IHC)", War Department, Washington, D.C., 1943

"Technical Manual TM9-2010 Multiple Caliber .50 Machine Gun Mounts M45, M45C, M45D, and M45F; Multiple Caliber .50 Machine Gun Trailer Mount M55; and Mount Trailer M20", Department of the Army, Washington, D.C., 4 December 1953

"Technical Manual TM9-2016 Ring Mounts M49, M49A1, M49A1C, and M49C; Truck Mounts M32, M36, M36A1, M36A2, M37, M37A1, M37A2, M37A3, M50, M56, M57, M58, M59, M60, and M61; Machine Gun Mounts M35C and M48; and Pedestal Truck Mounts M24A2, M24A3, M25, M31A1, and M31C

"Technical Manual TM9-2200 Small Arms, Light Field Mortars and 20mm Aircraft Guns", War Department, Washington, D.C., 11 October 1943

"Technical Manual TM9-2200 Small Arms Materiel and Associated Equipment", Department of the Army, Washington, D.C., 14 April 1949

"Technical Manual TM9-2200 Small Arms Materiel and Associated Equipment", Department of the Army, Washington, D.C., 9 October 1956

"Test of Half-Track Personnel Carriers M3E2" General Motors Proving Ground, Milford, Michigan, 8 February 1943

"Transportation Equipment and Related Problems" Summary Technical Report of Division 12, National Defense Research Committee, Volume 1, Washington, D.C., 1946

INDEX

Acceptances of Half-track Vehicles, 208
Allis-Chalmers Manufacturing Company, 174
American Ordnance Corporation, 157
Amphibian half-tracks, 172-174
Antiaircraft Vehicles, 122-157
  Combination Gun Motor Carriage M15E1, 135
  Combination Gun Motor Carriage M15A1, 135-137, 187, 192, 195, 221
  Multiple Gun Motor Carriage M13, 124-125, 142, 185, 205, 222
  Multiple Gun Motor Carriage M14, 126-127, 199, 223
  Multiple Gun Motor Carriage M15, 131-135, 137, 139, 187, 193, 199, 221
  M15 Special, 193
  Multiple Gun Motor Carriage M16, 124, 140-146, 149, 185, 193, 195-197, 222
    with bat wings, 196
  Multiple Gun Motor Carriage M16A1, 196-198, 222
  Multiple Gun Motor Carriage M16A2, 196-198
  Multiple Gun Motor Carriage M16B, 190
  Multiple Gun Motor Carriage M17, 140, 146-147, 199-200, 223
  40mm Gun Motor Carriage M34, 195, 207
  Multiple Gun Motor Carriage T1, 122, 148
  Multiple Gun Motor Carriage T1E1, 123
  Multiple Gun Motor Carriage T1E2, 123, 148
  Multiple Gun Motor Carriage T1E3, 128
  Multiple Gun Motor Carriage T1E4, 123-124
  Multiple Gun Motor Carriage T28, 128-131, 138, 152
  Multiple Gun Motor Carriage T28E1, 130-131, 185
  Multiple Gun Motor Carriage T37, 138-140
  Multiple Gun Motor Carriage T37E1, 138-140
  Multiple Gun Motor Carriage T58, 140
  Multiple Gun motor Carriage T60, 154-155
  Multiple Gun Motor Carriage T60E1, 156
  Twin 20mm Gun Motor Carriage T10, 148
  Twin 20mm Gun Motor Carriage T10E1, 149-151, 224
  40mm Gun Motor Carriage T54, 131, 154, 156
  40mm Gun Motor Carriage T54E1, 154. 156, 225
  40mm Gun Motor Carriage T59, 154-156
  40mm Gun Motor Carriage T59E1, 156
  Twin 40mm Gun Motor Carriage T68, 157
Anzio, Italy, 186
Armor
  Face hardened, 29, 46, 107
  homogeneous welded, 45-46
  sloped spaced for M3, 73
Autocar Company, 28, 52, 96-97, 130, 167
Bataan, 177
Battalions
  93rd Antitank, 97
  443rd Antiaircraft Automatic Weapons, Self-propelled, 130, 185

447th Antiaircraft Automatic Weapons, Self-propelled, 191
Bendix Aviation Corporation, 122
Bendix turret, 122-123, 128
Bowen & McLaughlin, 113, 196
Cape Gloucester, 193
Christie, J. Walter, 10
  half-track conversions, 10
Citroen-Kegresse, 11, 12
Crabb, Lieutenant Colonel Frederick, 40
Diamond T Motor Car Company, 27-28, 52, 109, 113, 170
Diesel engines, 45
Divisions
  2nd Armored, 40, 130, 188-189, 191, 204
Elco B-6 turret, 152
Engines
  American LaFrance V12, 16
  Buick Series 60, 26
  Cadillac V8, 12
  Continental R6572, 171
  Continental R975-C4, 172
  Ford V8, 20
  GMC 6 cylinder, 15-16
  GMC 6-71, 167
  Hercules 95 hp, 25
  Hercules DWXC, 45
  Hercules HXE, 19
  Hercules RXLD, 167, 170
  Hercules WXLC3, 26
  IHC Model RED-450-B, 50,65
  White 20A, 26
  White 24AX, 167
  White 140A, 26
  White 160A, 26, 28
  White 160AX, 38
Engineer half-tracks, 158
Four Wheel Drive Auto Company, 9
France, 187-188
Garford 3 ton truck, 9
General Motors Corporation, 17
Germany, 192
Great Britain, 199
Guadalcanal, 177
Guns
  20mm AN-M1 or AN-M2 (Hispano-Suiza), 148
  20mm Mark IV (Oerlikon), 148, 152, 229
  37mm Automatic M1A2, 128, 130, 193, 229
  37mm M3, 230
  40mm M1, 153-157, 161-162, 193, 231
  57mm M1, 106, 232
  75mm M1897A4, 97, 105, 234
  75mm M3, 105
  75mm T15, 105
  6 pounder Mark III, 106, 232
  6 pounder Mark V, 106, 232
Half-track Amphibian Cargo Carrier T32, 173
Half-track Cars, 12-14, 177
  M1, 12, 14
  M2, 27, 29-33, 39, 42-43, 45, 50, 52, 72, 80-81, 123, 128-129, 140, 148, 152, 163, 179, 204, 210
    with 37mm gun, 188
  M2A1, 29, 51-56, 192, 210

M2E1, 45
M2E4, 45
M2E5, 46
M2E6, 51
M3A2, 73-78, 211
M5A2, 73, 79, 213
M9, 46, 52, 72
M9A1, 51-52, 67-71, 199-200, 212
T1, 12
T1E1, 12-13, 15
T1E2, 12-13, 15
T1E3, 12-13
T2, 12
T6, 15
T14, 26-30
T16, 163-166, 226
T29, 72-73
T31, 72-73
Half-track Instrument Carrier T18, 154, 15
Half-track Litter Carrier T28, 174
Half-track Personnel Carriers, 25,
  M3, 27, 33-39, 44-45, 50, 72, 73, 80, 90, 97-98, 106, 112, 118, 121, 123, 129, 131, 140, 153, 157-159, 180, 183-184, 192-193, 204, 211
  M3A1, 51-52, 57-61, 72, 96, 105, 192, 211
  M3E1, 45
  M3E2, 46
  M5, 46-50, 72, 140, 199, 213
  M5A1, 51-52, 62-66, 72, 199, 213
  T7, 25-26
  T8, 27
Half-track Radio Carriers
  T17, 160
  T17E1, 160
  USMC Communications, 193
Half-track Trucks, 15-22
  Chevrolet 1 ton, 15
  Ford 1 1/2 ton, 15
  M2, 21
  T1, 15-17
  T2, 15
  T3, 16, 19
  T4, 16
  T4E1, 16
  T5, 17-18
  T5E1, 17
  T5E2, 17
  T6, 19
  T7, 19
  T8, 19
  T9, 20-21, 25, 28
  T9E1, 20-22, 28
  T9E2, 21-22
  T10, 21
  T14, 167
  T15, 167
  T16, 167, 170
  T17, 167-169
  T18, 167, 170
  T19, 167, 171
Harmon, Major General Earnest N., 40
Hauss, First Lieutenant Thomas., 40
Headlights
  Fixed, 40-41

Demountable, 40-41, 107, 112-113
Hinds, Colonel Sidney R., 40
Holt Tractor Company, 9
Houston, Donald E. , 40
Howitzers
    75mm M1A1, 118, 233
    105mm M2A1, 112-113, 236
    105mm M3, 235
Icks, Colonel Robert J., 97
Idler, spring loaded, 40
International Harvester Company, 45-46, 52,
    73, 146, 174, 199
Italy, 185-187
James Cunningham, Son and Company, 12,
    17
Jeffrey Quad, 10
    2 ton Model 4017, 10
Korea, 195-198
Lend-Lease Program, 52, 199
Lightweight Half-tracks, 174
Linn Manufacturing Company, WD-12, 16
Louisiana maneuvers, 177, 179
MacArthur, General of the Army Douglas, 177
Machine guns
    .30 caliber M1917A1, 29, 228
    .30 caliber M1919A4, 29, 33, 51, 72, 228
    .50 caliber M2, 29, 33, 51, 72, 98-100, 122-
    123, 126, 128, 130-131, 135, 137-140, 145-
    146, 152, 154, 160, 167, 228
Mack Manufacturing Corporation, 10, 161
    truck, 10
    Roadless, 10
    Half-track Chassis T3, 161, 163
    Half-track Truck T19, 171
    40 mm Gun Motor Carriage T1, 161-162
    105mm Howitzer Motor Carriage T34, 163
Marmon-Herrington DHT-5, 28
Marshall, General of the Army George C., 97
Marshall Islands, 193
Martin turret, 128
Maxson turret, 123,128, 140, 148
McKeen Motor Car Company, 9
Medical half-tracks, 158
Medaris, Major General John B., 190
Mine exploders, 158

Mine racks, 39
Mortars
    81mm M1, 80-81, 85, 90, 237
    4.2 inch M2, 96, 237
Mortar Carriers, 80-96
    81mm M4, 80-84, 96, 214
    81mm M4 Modified, 90, 189
    81mm M4A1, 85-89, 214
    81mm M21, 90-95, 215
    81mm T19, 90
    4.2 inch T21, 96
    4.2 inch T21E1, 96, 216
Nash Quad, 10
    2 ton Model 4017, 10
New Guinea, 194
Noble, Master Sergeant Gerry, 40
North Africa, 40, 184-186, 188
Packard 3 ton truck, 9
Patton, General George S. Jr., 108,185
Patton's half-track, 180
Philippines, 177, 193-194
    Luzon, 178, 194
QMC 5 ton BBW, 11
R. Hoe and Company, 148
Saipan, 193
Scott, Major General Charles L., 130
Scout Cars, 25
    M2A1, 25
    M3, 25
Self-propelled Artillery, 112-121
    75mm Howitzer Motor Carriage T30, 118-
    121, 205, 219
    105mm Howitzer Motor Carriage T19, 112-
    117, 182-183, 188, 220
    105mm Howitzer Motor Carriage T34, 163
    105mm Howitzer Motor Carriage T38, 121
Sicily, 186, 188
Signal Service Companies
    3132nd, 159
    3133rd, 160
Snow Tractors, 174
    M7, 174
    T26, 174
    T26E1, 174
    T26E2, 174

T26E3, 174
T26E4, 174
T27, 174
T27E1, 174
T27E2, 174
T27E3, 174
T27E4, 174
T27E5, 174
T29, 174
T29E1, 174
Sonic deception half-tracks, 159
Soviet Union, 146, 199
Spring loaded idler, 40
Tank Destroyers, 97-111
    75mm Gun Motor Carriage M3, 99, 101-
    106, 177, 181, 184, 188, 193, 199, 203, 218
    75mm Gun Motor Carriage M3A1, 104-105,
    218
    57mm Gun Motor Carriage T48, 106-111
    199-200, 217
    75mm Gun Motor Carriage T12, 97-100,
    104, 177-178, 181
    75mm Gun Motor Carriage T73, 105
Three-quarter track vehicles, 161-171
Tracks
    Band type, 15
    Goodrich, 39
    T6, 12
    T10, 15
    T14, 15, 17
    T17, 15
    T20E2, 20
    T21, 20
    T24E1, 21
Trackson drive, 19
Tucker Sno-Cat Company, 174
United Shoe Machinery Company, 138, 148
United States Marine Corps, 177, 193
Vehicle Data Sheets, 209
Weapon Data Sheets, 227
Welded armor, 45-46
White Motor Company, 26, 28, 52, 80, 90,
    121, 124, 142, 149, 167
W..L. Maxson Corporation, 123, 128, 140,
    148

www.ingramcontent.com/pod-product-compliance
Lightning Source LLC
Chambersburg PA
CBHW050040220326
41599CB00044B/7227